The Institute of British Geographers
Studies in Geography

Russia in the Modern World

D1550660

IBG STUDIES IN GEOGRAPHY

General Editor
Chris Philo

IBG Studies in Geography are a range of stimulating texts which critically summarize the latest developments across the entire field of geography. Intended for students around the world, the series is published by Blackwell Publishers on behalf of the Institute of British Geographers.

Published

Debt and Development
Stuart Corbridge

Service Industries in the World Economy
Peter Daniels

The Changing Geography of China
Frank Leeming

Russia in the Modern World
Denis J. B. Shaw

Critical Issues in Tourism
Gareth Shaw and Allan M. Williams

The European Community (second edition)
Allan M. Williams

In preparation

Geography and Gender
Liz Bondi

Population Geography
A. G. Champion, A. Findlay and E. Graham

Rural Geography
Paul Cloke

The Geography of Crime and Policing
Nick Fyfe

Fluvial Geomorphology
Keith Richards

The Sources and Uses of Energy
John Soussan

Retail Restructuring
Neil Wrigley

RUSSIA IN THE MODERN WORLD

A New Geography

Denis J. B. Shaw

350 Main Street, Malden, MA 02148-5018, USA
108 Cowley Road, Oxford OX4 1JF, UK
550 Swanston Street, Carlton South, Melbourne, Victoria 3053, Australia
Kurfürstendamm 57, 10707 Berlin, Germany

First published 1999 by Blackwell Publishing Ltd
Reprinted 2001, 2002

Library of Congress Cataloging-in-Publication Data has been applied for

ISBN 0–631–179038 (hbk)
ISBN 0–631–181342 (pbk)

A catalogue record for this title is available from the British Library.

Set in 10 on 12pt Plantin
by Best-set Typesetter Ltd., Hong Kong
Printed and bound in the United Kingdom
by MPG Books Ltd, Bodmin, Cornwall

For further information on
Blackwell Publishing, visit our website:
http://www.blackwellpublishing.com

For Andrea, Andrew and Hannah

Contents

List of Tables

List of Maps

The focus of this book is the changing economic, political and social geography of Russia since the fall of communism in 1991. Until now it has been customary for geographers writing about Russia to do so in the form in which it existed from 1917 to 1991 as the USSR, even though the Russian Federation was only one of fifteen Union republics. This was understandable given the highly centralized and Russocentric nature of the Soviet state. Since 1991, however, this approach has become increasingly anachronistic. After the fall of communism, each of the post-Soviet states began to follow its own path in the world, conscious of its own unique problems, culture and history. It is no longer possible to give each of these states adequate attention in a single geography text. It is time to give the new Russian Federation, as by far the biggest and most powerful of them, a text to itself.

Having made this point, it would of course be absurd to write as if Russia had only recently emerged on to the world scene or as if it stood alone. As this book is at pains to point out, the Russian state dates back to medieval times and eventually became one of the greatest empires in the world. Many would argue that the Soviet Union was an imperial successor to the empire which perished in 1917. This book is written in the belief that it is impossible to understand present-day Russia without knowing something of its long and remarkable history and especially of the history of the Soviet period. For that reason too, Russia must continue to be viewed, for the time being at least, in the context of its post-Soviet neighbours, particularly those in the Commonwealth of Independent States. Furthermore, because Russia exists in a globalizing world, it must also be viewed in a broader international context; hence the title of this book.

In the light of the fact that Russia can no longer be considered a superpower and potential rival to the United States for global supremacy, some may wonder whether the writing of geographies dealing solely with that part of the world can now be justified. Some may go further and question the value of any form of regional geography. My own view is that the writing of regional geographies is still an important exercise in geographical scholarship. The attempt to understand the human geography of a given portion of the globe in all the richness and complexity of its historical development, environmental diversity and spatial interconnections is surely still worth the effort, and is something which, arguably, only geographers can do. If nothing else, it can provide a check to some of the wilder flights of social theory, and provide empirical evidence for such theorizing. In Russia's case, I believe there to be additional reasons for the attempt. Despite the recent split, Russia is still the world's biggest state and continues to be one of the most important. Given its size, resource endowment and geopolitical situation, it is likely to continue to be influential in world affairs, sometimes controversially, no doubt. Moreover, Russia and its post-Soviet neighbours provide an important example of the attempt to develop an industrial system in a non-capitalist way and of the difficulties which beset the transition back to the market. Russia and its neighbours may well give even bullish proponents of the 'globalization of culture' thesis pause for thought.

Many people have helped with this book, either directly or indirectly. I should like to thank Sergey Artobolevsky, Andrew Bond, Tony French, Phil Hanson, Gregory Ioffe, Moya Flynn, Tatyana Nefedova, Bob North, Alexey Novikov, Judy Pallot, Phil Pryde, Matt Sagers, Douglas Sutherland and Andrey Treyvish. Kevin Burkhill and Anne Ankcorn of the School of Geography, University of Birmingham, drew many of the maps efficiently and professionally. My particular thanks go to Mike Bradshaw for his unfailing help and advice in numerous areas, to Jon Oldfield who drew some of the computer maps, and who eagerly assisted in so many other ways, and last but not least to my wife, Andrea, for her cartographic skills and constant and irreplaceable support. For mistakes and misapprehensions I thank only myself.

Denis J. B. Shaw
School of Geography
University of Birmingham

Acknowledgements

The author and publisher wish to thank the following for permission to reproduce the following material: M. J. Bradshaw and J. A. Palacin for figures 5.2, 5.3, 5.4, 5.5 and 5.6; Pennsylvania State University Press for the quote from Manuel Castells on page 269; V. H. Winston & Son Inc. for figure 8.1.

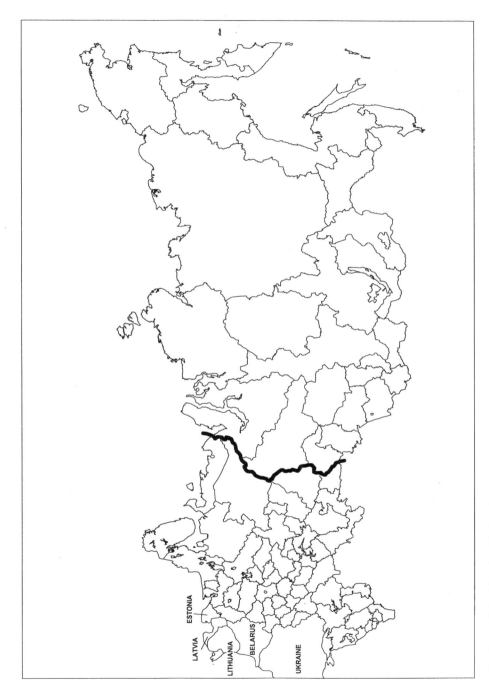

ESTONIA

LATVIA

LITHUANIA

BELARUS

UKRAINE

Regions of the Russian Federation

Key to the Regions of the Russian Federation

1. Aga-Buryat Autonomous Okrug
2. Adygeya Republic
3. Altay Kray
4. Amur Oblast
5. Arkhangel'sk Oblast
6. Astrakhan Oblast
7. Republic of Bashkortostan
8. Belgorod Oblast
9. Bryansk Oblast
10. Republic of Buryatia
11. Vladimir Oblast
12. Volgograd Oblast
13. Vologda Oblast
14. Voronezh Oblast
15. Republic of Gornyy Altay
16. Republic of Dagestan
17. Jewish (Yevreyskiy) Autonomous Oblast
18. Ivanovo Oblast
19. Ingush Republic
20. Irkutsk Oblast
21. Kabardino-Balkar Republic
22. Kaliningrad Oblast
23. Republic of Kalmykia-Khalmg-Tangch
24. Kaluga Oblast
25. Kamchatka Oblast
26. Karachayevo-Cherkess Republic
27. Republic of Karelia
28. Kemerovo Oblast
29. Kirov Oblast
30. Komi Republic
31. Komi-Permyak Autonomous Okrug
32. Koryak Autonomous Okrug
33. Kostroma Oblast
34. Krasnodar Kray
35. Krasnoyarsk Kray
36. Kurgan Oblast
37. Kursk Oblast
38. Leningrad Oblast
39. Lipetsk Oblast
40. Magadan Oblast
41. Republic of Mari-El (Mariyy El)
42. Republic of Mordovia (Mordvinian Republic)
43. Moscow (city)
44. Moscow Oblast
45. Murmansk Oblast
46. Nenets Autonomous Okrug
47. Nizhniy Novgorod (Nizhegorodskiy) Oblast
48. Novgorod Oblast
49. Novosibirsk Oblast
50. Omsk Oblast
51. Orenburg Oblast
52. Orel (Orlovskiy) Oblast
53. Penza Oblast
54. Perm' Oblast
55. Primorskiy Kray
56. Pskov Oblast
57. Rostov Oblast
58. Ryazan Oblast
59. Samara Oblast
60. St Petersburg
61. Saratov Oblast
62. Republic of Sakha (Yakutia)
63. Sakhalin Oblast
64. Sverdlovsk Oblast
65. Republic of Severnaya Osetia (North Osetia)
66. Smolensk Oblast
67. Stavropol' Kray
68. Taymyr (Dolgano-Nenets) Autonomous Okrug
69. Tambov Oblast
70. Republic of Tatarstan
71. Tver' Oblast
72. Tomsk Oblast
73. Republic of Tyva
74. Tula Oblast
75. Tyumen' Oblast
76. Udmurt Republic
77. Ul'yanovsk Oblast
78. Ust'-Orda Buryat Autonomous Okrug
79. Khabarovsk Kray
80. Republic of Khakasia
81. Khanty-Mansi Autonomous Okrug
82. Chelyabinsk Oblast
83. Chechen Republic
84. Chita Oblast
85. Chuvash Republic
86. Chukchi Autonomous Okrug (Chukotka)
87. Evenki Autonomous Okrug
88. Yamalo-Nenets Autonomous Okrug
89. Yaroslavl' Oblast

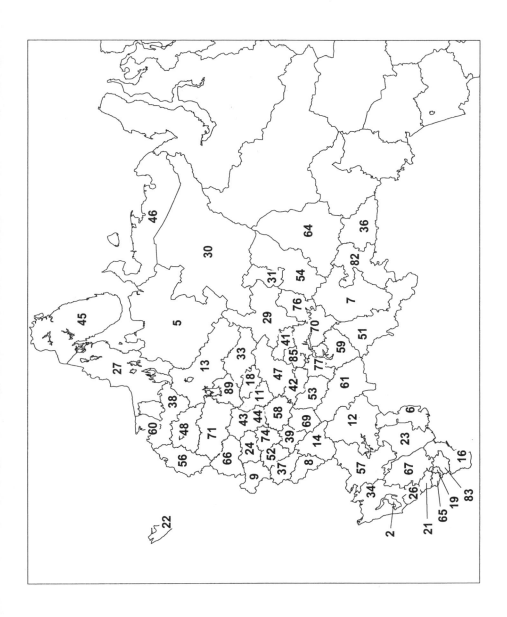

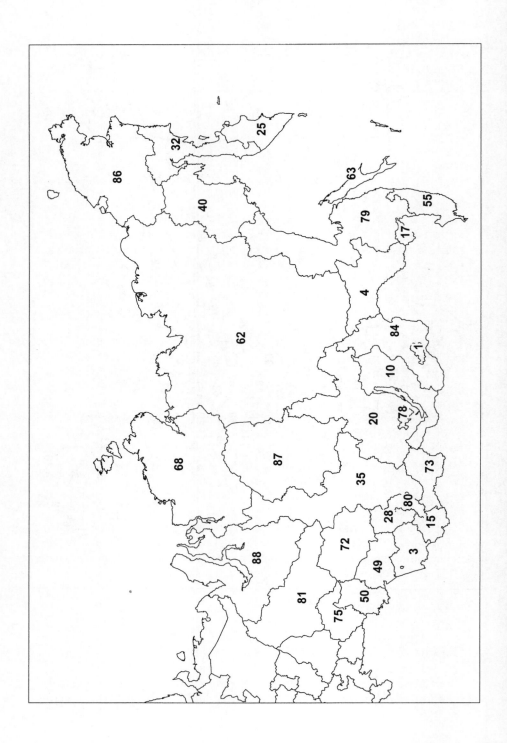

One

The Territorial and Imperial Heritage

The Russian Federation, or Russia as it is also officially styled, is the biggest of the fifteen newly independent states to have emerged from the ruins of the former Soviet Union in 1991. It is thus one of Europe's newest states. Oddly enough, however, it is also one of the oldest. Before 1991, the Russian Federation was part of the USSR or Soviet Union, a Russian-dominated state which had been established in the years after 1917 to replace the defunct Russian Empire. That empire in turn had grown gradually out of the small principality of Muscovy, which first appeared in medieval times. The purpose of this book is to examine the nature of the new Russia from a geographical perspective and to suggest that its problems cannot be understood without some knowledge of the long and eventful history which led up to them. An important theme is Russia's changing relationship with the world outside its frontiers. The present chapter will consider the territorial and environmental heritages which have been bequeathed to present-day Russia and how many of its problems relate to long-term development issues. Chapter 2 will look at the most recent, Soviet period and analyse its significance for Russia today. The ways in which the new Russia is now having to adjust to the modern world, and its prospects for the future, are the principal issues to be addressed in the chapters that follow.

A map of the Russian Federation will be found in figure 1.1. Figure 1.2 shows the location of Russia in relation to its other post-Soviet neighbours, while figure 1.3 depicts the territory of the former Soviet Union divided into economic regions, which are referred to throughout this book. A map showing Russia's physical features and regions will be found in figure 1.6.

Russia: The Territory and Its Acquisition

The Russian Federation contains over three-quarters of the territory of the former USSR. This means that, like the USSR, it is easily the largest

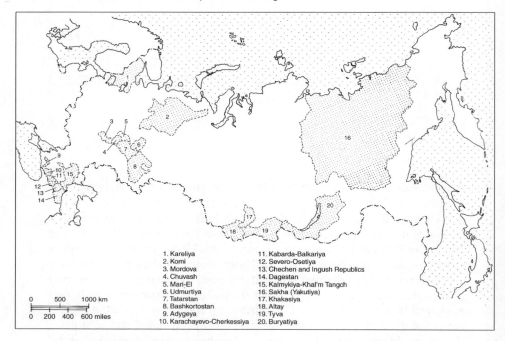

Figure 1.1 The Russian Federation and its constituent republics.

1. Kareliya
2. Komi
3. Mordova
4. Chuvash
5. Mari-El
6. Udmurtiya
7. Tatarstan
8. Bashkortostan
9. Adygeya
10. Karachayevo-Cherkessiya
11. Kabarda-Balkariya
12. Severo-Osetiya
13. Chechen and Ingush Republics
14. Dagestan
15. Kalmykiya-Khal'm Tangch
16. Sakha (Yakutiya)
17. Khakasiya
18. Altay
19. Tyva
20. Buryatiya

state in the world. At some 17.1 million square kilometres, Russia is 1.8 times the size of China, 1.7 times that of the United States and 70 times that of the United Kingdom. Its east–west extent is enormous, stretching from Kaliningrad on the Baltic to the Bering Strait in the Far East, and embracing eleven time zones. Its north–south axis is also impressive, as Russia reaches into the polar regions in the far north and extends southwards down to the semi-desert along the fringes of the Caspian Sea and to Mediterranean-type conditions on its fragment of Black Sea coast.

Before we consider the natural endowments of this territory and their significance for Russia and its future, it is worth dwelling at some length on how Russia came to be so large, since this is important to our later discussion of the country's geopolitical and ethnic problems. Russia's present-day borders were determined by the Bolsheviks or communists, who seized power in the Russian Revolution of 1917 and who later established the USSR. The territory of the USSR as a whole, in fact, very nearly corresponded with that of its predecessor, the Russian Empire. The latter had grown by a complex process of territorial aggrandizement from the mere 20,000 square kilometres which had constituted the state

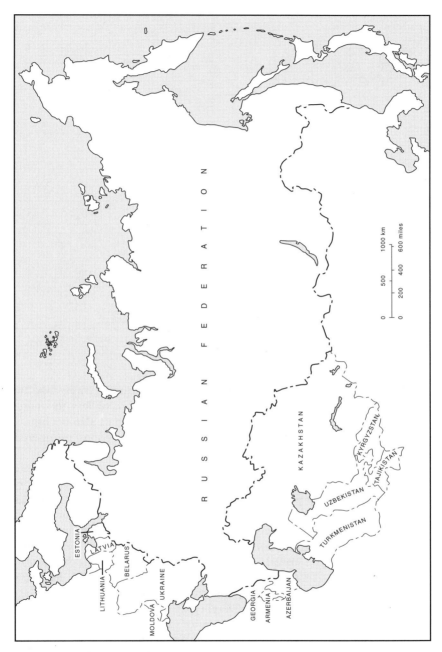

Figure 1.2 Russia and its post-Soviet neighbours.

1. North	11. Far East
2. Northwest	12. Donets-Dnepr
3. Centre	13. Southwest
4. Volga-Vyatka	14. South
5. Central Chernozem	15. Baltic
6. Volga	16. Transcaucasus
7. North Caucasus	17. Central Asia
8. Urals	18. Kazakhstan
9. West Siberia	19. Belorussia (Belarus')
10. East Siberia	20. Moldavia (Moldova)

---- USSR boundary

0   500   1000 km

0   200   400   600 miles

Figure 1.3   Economic regions of the former USSR.

of Muscovy at the beginning of the fourteenth century to more than 20 million by 1917. Like other European states at more or less the same period, therefore, Russia had founded an empire. It had also become a multi-ethnic state in the process. In 1917, ethnic Russians numbered fewer than half of the empire's subjects.

Later chapters of this book will show how the distribution of ethnic Russians across the territory of the former USSR is of enormous importance for the internal politics of present-day Russia and for its relations with its neighbours. In order to understand this distribution, however, it is necessary to know by what means the Russian Empire expanded and something about the migrations which accompanied it. This can be done by breaking down the historical process of expansion and settlement into a series of types. A useful device for this purpose is D. W. Meinig's 'macrogeography of Western imperialism' (Meinig, 1968). Meinig classified Western imperialism into several types, depending upon the different kinds of settlement and economic exploitation experienced in different regions. Unfortunately, Meinig's classification is not entirely suitable for the present purpose, partly because he was interested only in

European overseas imperialism, and partly because his scheme is rather generalized in scope. However, appropriately modified, it is helpful in dividing the Russian settlement experience into a series of separate types, corresponding to different regions of the former empire (figures 1.4 and 1.5).

The first part to be identified is the Russian 'homeland', the territory from which imperial expansion originated. This is less simple to define than might be imagined, simply because the imperial phase was preceded by a phase of 'internal' migration, territorial conquest and settlement, much as happened in Spain at the expense of the Moors, or in England/Britain at the expense of the Celtic peoples of the west. The Russian homeland is perhaps most convincingly defined as the territory occupied by Russian agricultural settlement at the time of the emergence of a unified Muscovite state (late fifteenth–early sixteenth centuries). So defined, this territory corresponds with a segment of European Russia stretching from Smolensk in the west to Nizhniy Novgorod in the east, and from Vologda and Velikiy Ustyug in the north to Tula in the south. It excludes most of the European north, where a minimal amount of agricultural settlement had occurred. It is overwhelmingly Russian in population and today falls entirely within the Russian Federation.

A second part can be identified with territories which are the historic homelands of peoples with a close ethnic and cultural affinity with the Russians, namely the Ukrainians and Belarusians. Territorially, this region corresponds with Belarus' and the northern and central parts of Ukraine. Before the thirteenth-century Mongol conquest of Russia, the Ukrainians and Belarusians developed in close association with the Russians, sharing much in the way of language, religion and other cultural traits. Thereafter, for several centuries, they were dominated by powers such as Lithuania and Poland, and only from the seventeenth century were they gradually annexed to Russia. This gradual absorption by Russia helps to explain the variations in culture across both Ukraine and Belarus' today. Thus, whereas the eastern parts of both states are culturally close to Russia (and after their annexation by Russia began to receive Russian migrants), their western parts (and especially those parts of Ukraine which were first annexed by the USSR only as recently as 1939–45) are much more distinctive. The Ukrainians and Belarusians played an important, if subordinate, role in building both the Russian Empire and its Soviet successor. Both Ukraine and Belarus' became independent states in 1991, but some Russian nationalists would like to see them reunited with Russia, both on account of their close historic links and because of the many ethnic Russians living there.

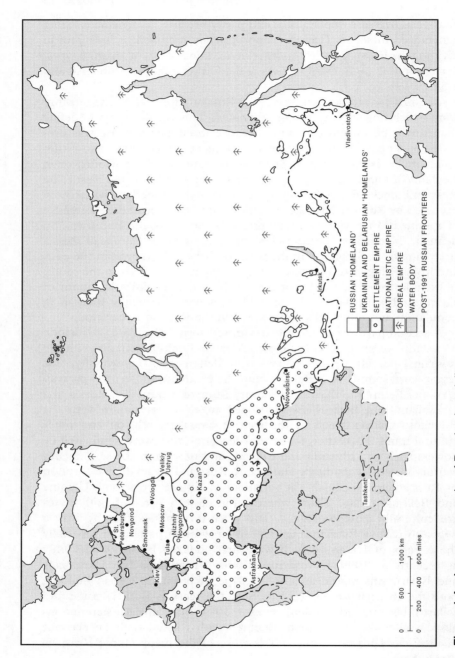

RUSSIAN 'HOMELAND'

UKRAINIAN AND BELARUSIAN 'HOMELANDS'

SETTLEMENT EMPIRE

NATIONALISTIC EMPIRE

BOREAL EMPIRE

WATER BODY

POST-1991 RUSSIAN FRONTIERS

Vladivostok

Irkutsk

Novosibirsk

Tashkent

St. Petersburg
Novgorod
Smolensk
Kiev
Tula
Nizhniy
Novgorod
Moscow
Vologda
Velikiy
Ustyug
Kazan'
Astrakhan'

0   500   1000 km

0   200   400   600 miles

Figure 1.4 A macrogeography of Russian imperialism.
Source: after Meinig (1968).

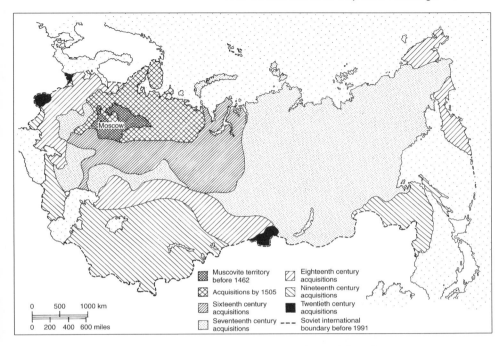

Figure 1.5   Territorial expansion of the Muscovite and Russian states.

The third region to be identified corresponds with what Meinig calls a 'boreal riverine empire'. This territory consists of the greater part of north European Russia falling within the boreal forest and tundra vegetational zones, and also most of Siberia and the Far East, apart from the very south (see figures 1.4 and 1.11). Meinig defines boreal riverine empires as those commercial empires which were built by Europeans in the boreal forests of North America and northern Eurasia for the exploitation of fur bearing animals. In their commercial character, they were similar to the contemporary sea empires of such powers as Portugal, Holland and England, but they differed from the latter empires in their overwhelming dependence on riverine communications. Initially, at least, there was a minimum of settlement. In Russia's case, this phase in its expansion began in the late medieval period, with penetration by the Russian city state of Novgorod of the northern and north-eastern European forests. The movement was accompanied by a certain amount of peasant and monastic colonization. Later, this region was annexed by Moscow. From the late sixteenth century, the Russian quest for furs spread across Siberia, reaching the Pacific in the middle of the following century. From here hunters and explorers moved southwards towards China and northeastwards towards the Bering Strait and Alaska.

Apart from having to pay the fur tribute, many of the northern native peoples, who are of varied ethnic origins (see chapter 3 and figure 3.2), were initially left to their own devices. Eventually, however, through peasant colonization in some instances, and the growing exploitation of minerals and timber in others, the number of Russians gradually increased. This was particularly true of the Soviet period, when exploitation of fuels and minerals brought large numbers of Russian settlers to the north. Today, therefore, the peoples of the north are usually easily outnumbered by the Russians within their own republics and autonomous regions (all of which now fall within the Russian Federation). However, many northern peoples retain a distinctive culture and sometimes a distinctive economy, so that they are conscious of their identity and often resentful of the way they have been treated by the white intruder.

Meinig's fourth region is the 'settler empire', by analogy with the white settlement of parts of the United States, Canada, Australia and other territories. Here the basic motive for expansion was permanent agricultural settlement rather than trade. Where Russian peasants came to settle permanently on northern lands, the boreal empire was gradually transformed into part of the settlement empire. But the greatest impact made by Russian agricultural settlement (outside the 'homeland') was across the southerly forest-steppe and steppe lands of the European territory (including southern Ukraine) and southern Siberia. It was also important in the mixed forests of the southern part of the Far East. These regions were largely settled by the Russians (and also to some degree by Ukrainians, Belarusians and some other groups) between the sixteenth and nineteenth centuries.

As was the case in the analogous regions of the United States, Canada and elsewhere, the indigenous populations of the settler empire, often nomadic and unamenable to European control, were frequently displaced, killed off, devastated by disease or otherwise reduced to insignificance. But, as with the white settlement of South Africa, parts of Latin America and other regions, the native populations were sometimes able to withstand European settlement, at least to the extent of retaining their identity and, occasionally, some of their land. The latter was most likely to occur where the indigenes were agriculturalists. Today, therefore, considerable numbers of non-Russians still live in some parts of the former settlement empire within the Russian Federation. Examples would include the Mari, Mordva, Chuvash and Udmurts of the middle Volga basin, the Bashkirs of the southern Urals and the Buryats of southern Siberia.

As a rule, white settlement in the settler empire had a mixed ethnic character, although the Russians usually predominated. The major exception was the steppelands of southern Ukraine, where Ukrainians played the principal role. Even here, however, many Russians also settled and their numbers were swelled from the late nineteenth century by migrants moving to take up jobs in the burgeoning industries of the Donbass coalfield and the Ukrainian south. The southern and eastern parts of the Ukrainian Republic still have a large Russian minority.

The final part of the old empire to be described corresponds with what Meinig termed the 'nationalistic empire'. Meinig used this term to denote the aggressive acquisition of overseas territory by Europeans, especially from the middle of the nineteenth century, when they were motivated by the search for national prestige, by international rivalries and by the need for raw materials and markets to serve their expanding industrial economies (the period often described by historians as the 'Age of Imperialism'). In fact this type of imperialism had antecedents both in Europe itself and overseas as incipient nation states vied with one another for territory and power. In Russia's case, because of the continental location of its territorial acquisitions, it is particularly difficult to distinguish between the 'nationalistic' imperial phase and a preceding phase in which nationalism was less apparent. What can be said is that this phase of imperialism, which began roughly in the eighteenth century, embraced territories peripheral to Russia proper which, because they were usually already well populated by non-Russians, experienced relatively little Russian settlement. Such territories include the former Baltic provinces (now the three Baltic states of Estonia, Latvia and Lithuania), the Transcaucasus (the states of Georgia, Armenia and Azerbaijan), parts of the North Caucasus now within the Russian Federation, most of Kazakhstan (with the probable exception of the north where there is much Russian settlement) and the four states of Central Asia (Uzbekistan, Kyrgyzstan, Tajikistan and Turkmenistan). Central Asia is probably the territory which best corresponds with Meinig's conception of 'nationalistic empire'. The southern part of the Far East, acquired from China in the mid-nineteenth century, was also annexed as a result of nationalistic imperialism (Bassin, 1983), but in view of its subsequent Russian settlement, it is probably best regarded as part of the settlement empire. It should be noted that the number of Russians living in the nationalistic empire was sometimes considerably increased as a result of policies pursued after 1917. This was most notably the case in Estonia, Latvia and northern Kazakhstan. It should also be stressed that most of the territory of the former nationalistic empire now lies outside the boundaries of the Russian Federation.

These suggested divisions of the former empire are, of course, based on a generalized and simplified picture of Russian expansion. The detailed situation was much more complex. How these divisions relate to the present-day boundaries of the Russian Federation and how they influence its geopolitical situation will be explored in subsequent chapters.

Russia's long history of territorial expansion has naturally interested a considerable number of scholars, some of whom have attempted to explain it by a single cause, such as an 'urge to the sea' (Kerner, 1946) or the innate migratory propensity of the Russians (Klyuchevsky, 1937, pp. 20–1). Some have even argued that it demonstrates the uniquely aggressive and expansionist character of the Russians. However, in view of the widely differing circumstances attending this long process of expansion, it is very doubtful whether such a complex phenomenon can be put down to a single cause (Bassin, 1988; Shaw, 1989). It is also surely unhelpful to cite certain peculiar character defects or traits in the Russians. Russian expansion, after all, is by no means unique. It took place at exactly the same time that other European powers were carving huge empires out of the world's islands and continents. The main difference is that the Russians were able to build their empire on territories contiguous to their homeland, and it is perhaps for this reason that they were able to keep them for so long.

Interestingly enough, in his discussion of the historical and political geography of imperialism, Peter Taylor fails to acknowledge Russia as having been one of Europe's imperial powers (Taylor, 1993, pp. 103–32). Perhaps this is because of the continental rather than overseas nature of its empire (although Alaska, a Russian possession until 1867, is an exception – see Gibson, 1976). Alternatively, world systems theory, which Taylor uses to explain European imperialism, postulates imperialism as a characteristic of core states within the world economy. Since, according to this theory, pre-1917 Russia was a semi-peripheral rather than a core state, it cannot have been an imperialistic power. But this seems too narrow a definition of 'imperialism'. Russian imperialism certainly had its peculiarities but, as noted above, it also had many similarities with that of other states, with many of the same forces at work (see Hunczak, 1974; Geyer, 1987; Rywkin, 1988).

The Natural Environment and Resources for Development

Now that something has been said about the story of Russian expansion, attention needs to be paid to the type of environment which the expanding Russian state encountered and the resources available for expansion both in the past and today.

The environment in which Russia originally developed and continues to exist is defined by a number of broad geographical parameters. The enormous size of the country has been commented on already. Russia lies astride the huge Eurasian landmass, the largest continous land surface on earth. The bulk of the territory is also situated relatively far to the north within the Northern Hemisphere. At 55° 45′ N, for example, Moscow is roughly at the latitude of Edinburgh in Scotland and north of Edmonton in Alberta, Canada. At 59° 55′ N, St Petersburg (formerly Leningrad) approximately parallels the Shetland Islands, north of Scotland, and the southern border of Canada's North West Territories. Mountainous terrain in the Caucasus, Central Asia, eastern Siberia and the Far East tends to shield Russian territory from warmer air masses which might otherwise penetrate from the Indian and Pacific Oceans (figure 1.6). By contrast, the lowlands of the north and west mean that much of Russia is open to incursions of cold Arctic air as well as to moist Atlantic influences coming from the west. The vastness of the territory ensures that much of it is far from the moderating climatic influences of the world's major oceans.

These broad factors mean that the Russian climate is a severe, continental one, with long, cold winters and relatively short, warm summers

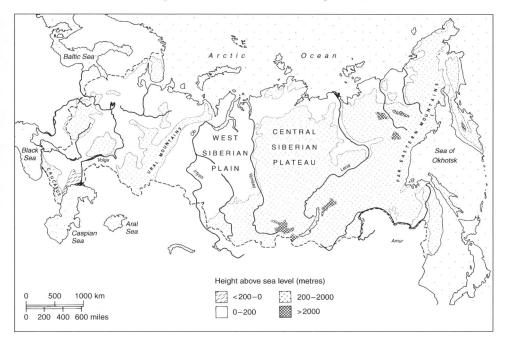

Figure 1.6 Russia: major physical features.

(table 1.1). The temperature range between summer and winter increases towards the east, with particularly cold winters being experienced in interior regions of northeast Asia (figures 1.7, 1.8). Similarly, since precipitation in these latitudes comes mostly from Atlantic air masses, it tends to decline as one travels to the east, until, in eastern parts of Siberia, Pacific influences begin to be felt (figure 1.9). Precipitation also falls away towards the north and, more particularly, towards the south. General aridity and drought are frequent problems of the southernmost parts of European Russia as well as southern Ukraine, but become even more of a problem as one approaches the semi-deserts of the Caspian Sea basin and Kazakhstan. Prevailing high pressure conditions in winter mean that precipitation maxima tend to occur in summer, which is not necessarily favourable for ripening crops. Practically the whole of the territory experiences winter snow cover, but precipitation patterns mean that this can be thin and powdery in some agricultural areas, a problem for wintering crops when temperatures are low.

Patterns of soils and natural vegetation are responses to these broad climatic factors. As figures 1.10 and 1.11 show, both soil and vegetation zones tend to run west–east across Russia in broad belts. Thus, across the north, where generally cool or cold conditions means that annual

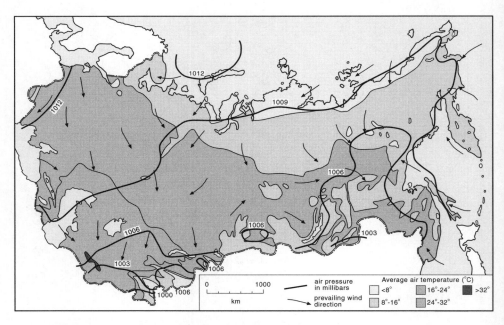

Figure 1.7   Former USSR: mean surface air temperatures, July.

Table 1.1 Climatic indicators for selected Russian cities

	Lat.	Long.	Elevation (m)	Mean temp. Jan. (°C)	Mean temp. July (°C)	Mean annual precip. (mm)	Frost-free days (average)	Snow depth max. (cm)[a]	Mean annual sunshine (hrs)
Arkhangel'sk	64° 30'N	40° 39'E	3.9	−12.6	15.6	539	119	62	1,576
Astrakhan	46° 16'N	48° 02'E	18	−6.9	25.1	190.1	187	–	2,441
Dudinka	69° 24'N	86° 10'E	20	−29.5	12.0	267	77	46	1,518
Irkutsk	52° 16'N	104° 79'E	468	−20.9	17.5	458	94	28	2,046
Kaliningrad	54° 42'N	20° 37'E	27	−2.7	17.3	698	182	73	1,786
Krasnodar	45° 02'N	39° 09'E	33	−2.1	22.9	640	186	42	2,146
Moscow	55° 45'N	37° 34'E	156	−10.3	17.8	575	141	52	1,597
Novosibirsk	55° 02'N	82° 54'E	162	−19	18.7	425	120	34	2,041
Okhotsk	59° 22'N	143° 12'E	6	−24.5	12.9	378	107	45	2,010
Perm'	57° 57'N	56° 13'E	169.7	−15.4	18.0	570	118	76	1,799
St Petersburg	59° 58'N	30° 18'E	4	−7.5	11.4	559	159	32	1,563
Salekhard	66° 32'N	66° 32'E	35	−24.4	13.8	464	94	59	1,512
Sochi	43° 35'N	39° 43'E	31	5.7	22.7	1,356	289	6	2,253
Verkhoyansk	67° 33'N	133° 23'E	137	−48.9	15.3	155	69	25	1,953
Vladivostok	43° 07'N	131° 54'E	138	−14.7	17.5	721	187	18	2,131

Note: [a] Average maximum snow cover for 10 days or more per annum.
Source: Lydolph (1977, pp. 363–427).

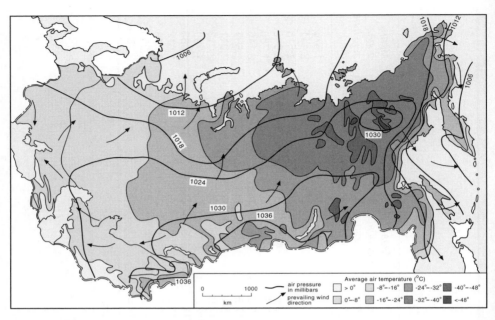

Figure 1.8 Former USSR: mean surface air temperatures, January.

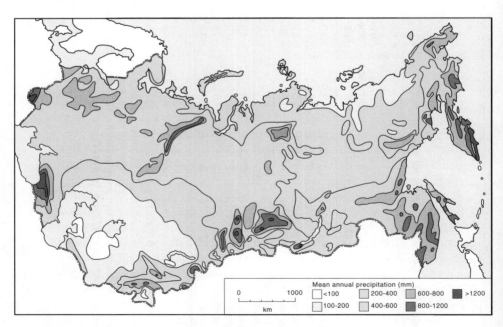

Figure 1.9 Former USSR: mean annual precipitation.

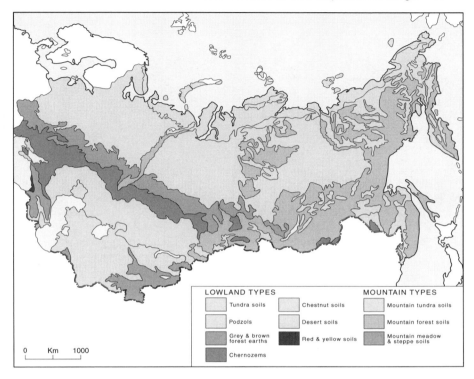

LOWLAND TYPES

- Tundra soils
- Podzols
- Grey & brown forest earths
- Chernozems
- Chestnut soils
- Desert soils
- Red & yellow soils

MOUNTAIN TYPES

- Mountain tundra soils
- Mountain forest soils
- Mountain meadow & steppe soils

0   Km   1000

Figure 1.10   Former USSR: soils.

precipitation tends to exceed evaporation, soils are acidic and the vegetation predominantly coniferous forest or, where colder, tundra. Across the south, petering out in eastern Siberia, run belts of forest-steppe and steppe grassland, deteriorating southwards (or southeastwards in European Russia) into semi-desert. In these regions evaporation exceeds precipitation and soils tend to be alkaline.

Historically, the soil and vegetation zones had a significant influence on Russian development. Thus the central part of what has been described earlier as the Russian 'homeland' corresponds with the mixed forest zone of European Russia. Here arose the city of Moscow and the historic Muscovite state in the late medieval period, in a region relatively well protected by forests from the nomadic warriors who roamed the open grassy steppelands in the drier conditions to the south. While the moderately leached podzols and grey forest soils of the mixed forest zone are not particularly rich, Russian peasants found cultivation fairly easy among the mixed stands of coniferous and deciduous trees, except where

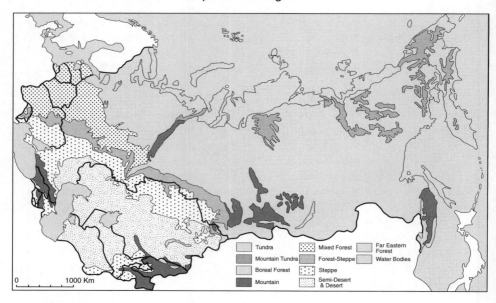

Figure 1.11   Former USSR: natural vegetation.

the surface had been strongly affected by glacial erosion or deposition. Further west towards the Baltic, moisture conditions improve, but these regions were annexed by Russia only in the seventeenth and eighteenth centuries (figure 1.5).

As noted already, Russian penetration of the boreal forests to the north began in medieval times, mainly in the quest for furs. To the east, across the Urals, the boreal forest belt becomes broader and the mixed forest disappears altogether. Russian penetration of this region, Meinig's boreal riverine empire, began, as we have seen, in the sixteenth century and continued down to the eighteenth. However, despite the presence of numerous natural resources, agriculture and settlement were everywhere restricted by the highly acidic soils, poor drainage, a generally severe climate and, in eastern Siberia and the Far East, mountainous terrain. However, in the Soviet period such disadvantages were partially redeemed in the eyes of government and of many migrants by the discovery of oil, gas and other riches. In the tundra to the north, the natural environment is even more severe, but once again the recent discovery of energy resources has led to development in some regions.

Russian settlement of the steppe grasslands lying to the south and southeast of the 'homeland' region was long hindered by the warlike nomads who roamed the area. Only in the seventeenth and eighteenth centuries was the Russian state strong enough to subdue the nomads and encourage the beginnings of agricultural settlement. Because of the predominance of fertile black earths and similar soils, the region proved particularly suitable for cultivation and was gradually converted into Russia's 'breadbasket'. The natural steppe grasslands thus virtually disappeared. The major natural barrier to cultivation is drought, which worsens towards the south and southeast. Because of the prevailing west–east atmospheric circulation patterns, the western parts of the steppe belt in Ukraine have more favourable moisture conditions than those further to the east, such as in southwest Siberia. Yet, in the context of Siberia itself, the latter region is the most significant for both agriculture and human settlement (see figures 7.1, 7.2).

In building their empire, the Russians also encountered other types of natural environment: for example, the Mediterranean environment of the Black Sea coast of Crimea, the 'subtropical' lands of the Transcaucasus and the semi-deserts and deserts of Central Asia. All these now lie in Russia's neighbouring republics. Mention should also be made of the mixed forest zone of Primorskiy Kray in the southern Far East, in the region annexed from China in the mid-nineteenth century. This area, still in Russia, contains plant species peculiar to eastern Asia and is agriculturally important. It is isolated from the steppelands of southwestern Siberia by the mountains of East Siberia and the Far East, with occasional intermontane basins having forest-steppe and steppe characteristics.

The 1991 break-up of the Soviet Union has greatly affected Russia's access to natural resources. Thus, in the late 1980s, with 51 per cent of the USSR's total population, the Russian Federation produced only 46 per cent of its agricultural output. The loss of the productive mixed forest, forest-steppe and steppe lands of the Baltic states, Belarus', Ukraine and Moldova, together with more specialized kinds of production in such regions as Transcaucasus and Central Asia, poses problems for government. This is because, for the geographical reasons just described, Russia's agricultural resources are less productive than those further to the west.

Equally, however, since Russia contains the vast and largely undeveloped regions of the north, Siberia and the Far East, its endowment of non-agricultural resources is often far superior to those found in the well populated and industrially developed western republics. As later chapters will show, Russia's reserves of oil, gas, coal, timber, water and other resources are huge, if only too often remote from the major centres of population and economic activity. Since the other post-Soviet republics have generally depended on Russia for a range of natural resources in the past, the latter seems well placed to exercise its economic influence over some of them in the future.

Russian Development: the Long-term Influences

Given Russia's long and successful history of territorial expansion, and its access to many kinds of resources, it might be thought that its development potential is a formidable one. Many people have argued this way. Early in the twentieth century, for example, the eminent British geographer Halford Mackinder maintained that whichever power occupied the centre of the huge Eurasian landmass, or 'geographical pivot' as he called it, not only had a unique defensive position and resource endowment but was well placed to dominate much of the rest of the world (Mackinder, 1904, 1919). Later, under the influence of the Cold War, Western powers feared the military might and apparent technological capabilities of their gigantic neighbour to the east. Yet the fact remains that throughout much of its history Russia has lagged economically behind the West, and this problem remains to haunt it today. The aim of this section is to suggest some of the long-term factors which lie behind this apparent economic backwardness. The rest of this chapter and the one which follows will then examine the development experience both before and after 1917 as a background to the problems which Russia faces now.

That the natural environment poses many problems for Russian development has been suggested already. Agricultural yields, for example, are and probably always have been well below those achieved in the West. The actual spatial extent of agriculture is severely limited, as noted above (see figure 7.2). Much of the north and east is far too cold for successful farming, while mountainous terrain characterizes many regions of East Siberia and the Far East. To the east of the Urals, in fact, arable farming is confined to a narrow strip sandwiched between the forests and swamps to the north and the Kazakh frontier and mountains to the south, with its apex towards Lake Baykal. West of the Urals, the most promising conditions are to be found in the centre and south, but even here things are often far from ideal. The mixed forest zone, for example, often suffers from acidic soils (grey forest soils or podzols), the after-effects of glaciation (swamps, boulders, stony soils), the shortness of the growing season (130 days around Moscow), the possibility of late spring or early autumn frosts and other problems (White, 1987, pp. 47–8). Further south, as we have seen, conditions improve, with more fertile soils and a longer growing season. But even here there are many difficulties: drought, soil erosion, insect pests, disease and, especially in the past, fuel shortages.

In traditional Russia, where agriculture was the basis of the economy, such problems meant that a considerable amount of time had to be devoted simply to the business of subsistence. There was therefore always a limited surplus available to meet the needs of the non-agricultural population and the demands of the state. This is still a problem today. However, it would be a mistake to place all the blame for the problems of agriculture on the natural environment. Agricultural technique and organization, conservative social attitudes and institutions such as serfdom arguably held back development to a greater or lesser degree in traditional Russia. In the modern period, as suggested in later chapters, Soviet agricultural policy had an extremely detrimental effect on the countryside.

A related problem which helped to hinder both agriculture and industry in traditional Russia was the lack of a market. This reflected not only the agricultural and social problems just mentioned but also other factors, like poor communications and the relatively low population density. The vast distances were difficult and expensive to cross by traditional means of transport and, although good use was made of the navigable rivers, these were affected by the long winter freeze, as indeed were many of the ports. Water-based communications were also hindered by the need for portages between river basins and around the occasional rapids. There were many cases where there was no alternative to using the expensive overland routes (although winter sledging was relatively cheap). Improvement of communications was delayed by both expense and conservative attitudes. Not until well into the second half of the nineteenth century did Russia begin to develop an adequate railway network, well behind the other major European powers.

In a classic article written many years ago, Alexander Baykov pointed out that Russia has been particularly unfortunate, not in the availability of industrial resources but in their distribution (Baykov, 1954). Thus Peter the Great, in seeking to develop the iron industry early in the eighteenth century, was forced to turn to the iron ores of the Urals, hundreds of miles to the east of Moscow. Later, in the industrialization drive of the late nineteenth century, the new iron and steel industries arose on the Donbass coalfield and Krivoy Rog ore field far to the south in the southern Ukraine (see figure 1.13). Only a modern railway system, as developed in the second half of the nineteenth century, could counteract such distances, and even then the expense was considerable. Access to resources has remained a problem in the recent period. In the Soviet era the continuing industrialization of the European territory of the

USSR made increasing demands on the resources of the east. Only through considerable investment in pipelines and other transportation systems was it possible to keep the European industries supplied with fuel and essential raw materials.

A further factor which has served to hinder Russian development has been its frequent isolation from the outside world. Russia's territorial expansion turned it into a huge, transcontinental country spanning Europe and Asia (figure 1.5). It is arguable that this expansion tended to exacerbate its marginal position relative to western Europe and the gradually developing world economy, sometimes turning its eyes away from the west and towards the east. In fact, relative isolation, at least from western Europe, has a long history. Geography, and also rivalry with such powerful neighbours as Poland–Lithuania, Sweden and the nomads of the southern steppe, denied Russia easy access to the Baltic and Black Seas for many centuries and tended to exaggerate its continental character (figure 1.12). When Richard Chancellor pioneered English trading contacts with Muscovy in 1553, his route took him far to the north around the North Cape and thence along the inhospitable Murman coast down to the White Sea. This for long remained Russia's easiest means of contact with the countries of western Europe, but it was extremely circuitous. In an age of difficult and expensive overland transport, the lack of any easy access to the sea was a distinct disadvantage. Only with Peter the Great's conquest of the Baltic coastline at the beginning of the eighteenth century (his great achievement here was the founding of his new capital, St Petersburg), and the acquisition of the Black Sea shore later in the same century, did the situation improve. Even then, Russia continued to feel insecure about its access to the world's oceans, especially given the narrowness of the exits from both Baltic and Black Seas and the fact that they were controlled by foreign powers.

In addition to the facts of geography, Russia's relative isolation was fostered by its cultural heritage. Christianity, for example, came to Russia in the tenth century (AD 988) in its Byzantine or Eastern Orthodox form. Geography is again implicated here, since medieval trading routes tended to follow the southward-flowing rivers to the Black Sea and thence to Constantinople (Byzantium), rather than overland routes towards the west. The result was partially to insulate the Russians against influences coming from the Catholic West, and caused them to regard such influences with deep suspicion. This attitude was reinforced in the thirteenth century, when Russia was invaded by the Mongol-Tatars, who were soon converted to Islam. Russia remained subservient to the Tatar khans (princes) for at least two centuries. Thus while the Belarusians and Ukrainians to the west now fell under the power of

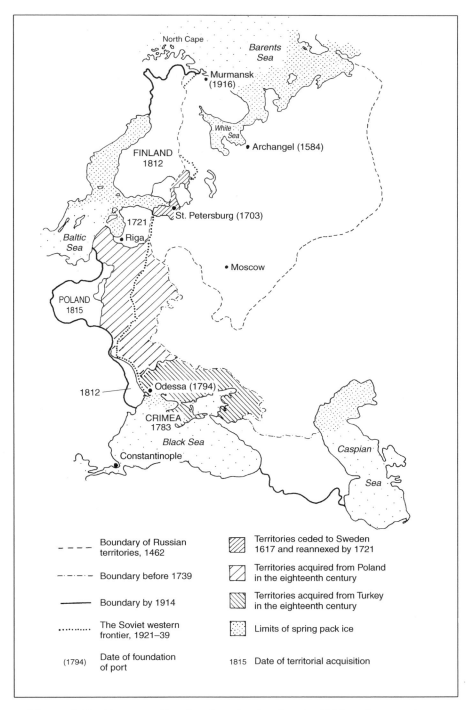

North Cape

Barents
Sea

Murmansk
(1916)

White
Sea

Archangel (1584)

FINLAND
1812

St. Petersburg (1703)

1721

Baltic
Sea

Riga

POLAND
1815

Moscow

1812

Odessa (1794)

CRIMEA
1783

Black Sea

Caspian

Constantinople

Sea

– – – – Boundary of Russian territories, 1462	▨	Territories ceded to Sweden 1617 and reannexed by 1721
– · – · – Boundary before 1739	◿	Territories acquired from Poland in the eighteenth century
——— Boundary by 1914	◹	Territories acquired from Turkey in the eighteenth century
·········· The Soviet western frontier, 1921–39	⬚	Limits of spring pack ice
(1794) Date of foundation of port	1815	Date of territorial acquisition

Figure 1.12 European Russia: changing frontiers to 1939 and access to the sea.

Lithuania and Poland, and were hence more open to European influences, the rise of the Muscovite state to the east took place in circumstances which cut it off from the rest of Europe. This was to have some peculiar effects. For example, the fall of Constantinople to the Muslim Turks in 1453 suggested to the Russian tsars that they were now the only true Orthodox sovereigns, whose sacred duty it was to uphold Orthodox Christian truth in the face of its many enemies. Inevitably, this made them and their people less than sympathetic to strange, new ideas coming from abroad. Until Peter the Great's reign (1682–1725), for example, foreign merchants residing in Moscow were required to live in a special suburb where they could practise their 'heretical' religions, in the words of contemporary documents, 'far away from the churches of God' (see Kochan and Abraham, 1983, pp. 96–7).

Russians also tended to be suspicious of outsiders because those outsiders so often came to Russia as its enemies. To the west and south of the European territory in particular, the open frontiers and the absence of any natural barriers have been a constant temptation to Russia's foes. The problems posed by the steppe nomads have been mentioned already. The most warlike were the Tatars, whose raids and incursions ended only in the eighteenth century. The need for extensive and sophisticated defensive measures along the southern frontier was a constant drain on Russia's resources (Shaw, 1983). Even more menacing in the longer term were the various states and peoples to Russia's west: the Poles, the Swedes, the Germans and others. From at least the thirteenth century, the western frontier suggested the dread possibility of invasion and conquest by one of the more powerful western peoples. Actual invasion occurred on numerous occasions, the last one, in 1941–5, accounting for many millions of Soviet lives.

The military needs of the Russian state (reflecting, of course, not only its defensive requirements but also its ambitions for expansion) were an ever-present reality. As one historian has written of the sixteenth and seventeenth centuries, 'War was the major preoccupation of the Muscovite state' (Hellie, 1971, p. 21). The state's military needs were a heavy burden on the Russian people. Not only were there immense territories to defend (space could also be an aid to defence, as Napoleon found to his cost in 1812 and Hitler in 1941–5) but Russia was far from wealthy, as suggested above. Immense efforts of organization and coercion were therefore needed to ensure that the available resources were mobilized in the required manner.

It seems probable that the open frontier and the military burden helped to foster the centralization of the Russian state. The autocratic powers of the tsars were legendary, particularly among visitors from the West. Thus Sigismund von Herberstein, who visited the court of Tsar

Vasiliy III in the early sixteenth century, wrote that the tsar 'surpasses all other kings and princes in the power he has and uses over his own people. – He holds one and all in the same subjection' (von Herberstein, 1966). Whether such an account of the tsars' authority is accurate is a matter for dispute. But it was the case that the tsars attempted to organize the whole of society to serve the interests of the state (and in the first instance, its defensive interests) and that Russia failed to evolve the autonomous institutions and social estates that were a characteristic of much of Europe. Unlike many of the Western powers, which, from about the sixteenth century, drew considerable wealth from participation in an expanding mercantile economy, traditional Russia had to derive most of its resources for defence and development from within its own frontiers. The source of much of that wealth was agriculture, and the indispensable prerequisite for agriculture was a peasant population. Unfortunately, Russia was for long a sparsely populated country and its population highly mobile. One of the 'achievements' of Muscovite centralization from the sixteenth century was to tie down the rural population through enserfment to its place of residence, where it could be taxed and made to support the landlord class. The latter was in turn obliged to serve the state. In western Europe, of course, serfdom had largely disappeared by this time. Its enforcement in Russia, and the attempted regimentation of the rest of Russian society, were reflections of the very different economic and political conditions existing there. By the nineteenth century the regimentation was eased, and serfdom was finally abolished in 1861. But the sense that the state was omnipresent and all-powerful was never entirely eliminated.

Yet there is an enormous difference between the apparent power of a state and its actual ability to control events, and nowhere was this more obvious than in pre-revolutionary Russia. Autocracy did not necessarily imply a strong state. According to John Keep, the opposite was true: the reality was that traditional Russia was a poor and undergoverned realm, and the posturings of its autocratic rulers were actually only tokens of the very real limits to their powers (Keep, 1972). The demands for regimentation and obedience were confessions of inevitable frustration. Here once again, territory and space were decisive, for outlying regions, and the vast distances of rural Russia, were particularly difficult to monitor and control. As will be seen later in this book, this problem is still very much in evidence today.

Only from the time of Peter the Great was a serious effort made to open Russia up to the outside world and to end its centuries of isolation from Europe. Even then, Peter's successors varied in their enthusiasm for this goal. After 1917, Russia shut itself away once again, claiming the status of the world's 'first socialist state', with a mission to save the rest of the globe from capitalism. Only recently has the process of opening up been resumed. This vacillating picture is described in the section and chapter to follow.

Economic and Social Development to 1917

Immanuel Wallerstein has argued that the fifteenth century witnessed the beginnings of what he calls a capitalist world economy (Wallerstein, 1974). This economy was initially based on western Europe but gradually expanded across the Atlantic and beyond. At first, according to Wallerstein, Russia did not form part of this growing world economy. Yet the Russian state had to exist in this newly dynamic world, and it was clearly necessary, if it was to retain its political integrity, to participate in it.

Russia was certainly one of the first countries to become self-conscious about its technical and economic backwardness relative to the developing West. In Peter the Great's time it was the country's military backwardness which initially inspired concern and caused him to modernize his armed forces (including, effectively for the first time in Russian history, the building of a navy). Peter also introduced a whole host of other social and economic reforms. Several subsequent rulers have pursued similar policies of modernization. Their motives have been mixed. In addition to the ever-pressing need for defence, there was Russia's growing ambition as one of Europe's great powers, a status it had clearly acquired by the latter half of the eighteenth century. And there was the question of prestige, the desire to be seen as modern, like other European states. As the nineteenth century historian V. O. Klyuchevskiy wrote: 'From Peter, hardly daring to consider themselves people and still not considering themselves proper Europeans, Russians under Catherine [the Great] felt themselves to be not only people, but almost the first people of Europe' (quoted by Dukes, 1982, p. 159). And in much the same vein, the Soviet leader Nikita Khrushchev wrote in his *Memoirs*: 'It's no small thing that we have lived to see the day when the Soviet Union is considered to be, in terms of its economic might, the second most powerful country in the world' (Talbott, 1974, p. 529).

Such pride in achievement has often carried as its mirror image nagging doubts about how real that achievement is, and about how powerful Russia has become by comparison with other states. There have also been worries about the nature of Russia's relations with foreign powers and about how positive or negative are foreign influences upon Russia's society and people. Just as occasionally happens in developing countries today, voices have been raised in Russia casting doubt on the appropriateness of the Western model of progress to Russian circumstances and advocating the idea that Russia should attempt to find its own, unique, development path. For reasons such as this, some scholars have argued that pre-revolutionary Russia should be considered the first of the 'developing nations' (Shanin, 1985). In Russia of the nineteenth century, the doubts gave way to sharp differences of opinion between, on the one hand, the Westernizers, who admired the West, or at least its technical achievements, and wanted Russia to develop along Western lines, and, on the other hand, the Slavophils, conscious of Russia's distinctiveness, often suspicious of foreign influences and alarmed lest those influences should destroy their country's cultural heritage. In some ways, Peter the Great was the first of the thoroughgoing Westernizers, and in his own day his policies were opposed by the forces of Muscovite xenophobia and conservatism. Later, during the reign of Catherine the Great (1762–96), who was herself of German origin, some writers and intellectuals began to pour scorn on the adopted French manners and language of the court and members of the higher nobility.

In the nineteenth century, Russian intellectuals began to express themselves more forcibly on either side. One of the most famous of the Westernizers, Vissarion Belinskiy, argued that because Peter the Great 'carried out in thirty years a task that needs centuries' – namely, the beginnings of the Europeanization of the Russian people – he is to be regarded as 'a giant among giants, a genius among geniuses, a king among kings': 'he is akin and equal to no-one but himself'. 'We want to be Russians', declared Belinskiy, 'in the European spirit' (Kohn, 1962, p. 128). Entirely contrary were the sentiments of Ivan Aksakov, a leading political publicist. In a memorable address following the assassination of Tsar Alexander II in 1881, Aksakov denounced the murder as an attack upon 'the autocratic power bestowed upon the Emperor by the country itself' and as the 'logical, extreme expression of that Westernism which, since the time of Peter the Great, has demoralized both our government and our society, and has already marred all the spiritual manifestations of our national life' (Kohn, 1962, p. 113).

Despite the undoubted intensity of this dispute, it was carried out against the background of a Russia which changed but slowly until almost the very end of the tsarist period. Peter's achievements were many, but some were undone after his death and the economic transformation he set in train atrophied for several decades. From about the middle of the eighteenth century, the Russian economy showed more signs of life: a growing population, an increase in international trade, more domestic market centres, a greater amount of land devoted to agriculture, a rather more vigorous life in towns and cities. But until the outbreak of the Crimean War (1853–6), industrialization was slow as the government viewed with alarm the social tensions and distresses it was bringing to other countries. Russia's disastrous defeat in the Crimean campaign, however, demonstrated just how relatively backward its military forces and economy had now become. This in turn spurred the government on towards a series of social reforms, beginning with the Emancipation of the Serfs in 1861, and towards policies for economic growth. Yet not until the 1880s was a concerted effort made to hurry the process of industrial modernization. Even after the Emancipation, the great majority of the population remained poor peasants. It was the continuing and great discrepancy between the minority of urban dwellers (and even smaller numbers of educated and well-to-do people) and the great majority of the peasants that helped to fuel doubts about the relevance of western Europe's experience of economic development to Russia's case.

Russia's relative poverty and its lack of a significant entrepreneurial class naturally constituted major barriers to development. Also important by the middle of the nineteenth century was potential competition from the other, more developed capitalist powers, such as Britain and Germany, whose modern and efficient industries threatened to strangle Russian industry at birth. For reasons such as these, Alexander Gerschenkron argued that Russian development was forced to rely heavily upon the state, which provided the necessary substitutions for the private capital and entrepreneurship that had characterized Western and, in the first instance, British industrial development (Gerschenkron, 1962). Gerschenkron and, more recently, Dieter Senghaas have tried to place Russian experience within a general model of industrial development for Europe as a whole (Senghaas, 1985). The later European countries embarked upon industrialization, they argue, the more difficult it was to achieve and the more important the role of the state became. Thus the Russian experience was by no means aberrant; nor was it simply the product of Russian centralization and absolutism. By the early years of the twentieth century, private capitalism was already playing a more important role. It is also important to note the opinion of some historians that the state's role in Russian industrialization was not always a positive one, often hindering as well as encouraging industrial development.

Senghaas describes pre-revolutionary Russian development as an example of 'state capitalism' (Senghaas, 1985). Russian economic policy, particularly as practised under Sergey Witte, Minister of Finance from 1892 to 1903, has been described by B. H. Sumner (1944, pp. 323 ff.). Witte's policies of industrialization included financial, monetary and legal reform, the encouragement of foreign capital investment, tariff protection and railway building. Industrial development made impressive strides during the three decades or so prior to the First World War, as described in tables 1.2 to 1.4.

If, in accordance with Wallerstein's world system theory, Russia had joined the world capitalist economy at some point in the eighteenth century, it was playing an important if semi-peripheral role by the outbreak of the First World War (Wallerstein, 1979, 1980). According to the theory, a 'semi-peripheral' status implies a position of dependency relative to one or more of the advanced 'core' countries, such as those of western Europe and North America, but one in which the semi-peripheral country itself invests in and reduces to dependency parts of the periphery (in the case of Russia, the periphery would at that time have embraced Siberia, Central Asia, possibly the Transcaucasus, parts of China and other regions). Thus, on the one hand, Russia was the target of a good deal of Western investment, although after 1900 actual ownership of Russian firms tended to be vested in the Russians

Table 1.2 Russian industrial output (manufacturing and mining), 1860–1913 (1900 = 100)

1860	13.9	1896	72.9	1905	98.2
1870	17.1	1897	77.8	1906	111.7
1880	28.2	1898	85.5	1907	116.9
1890	50.7	1899	95.3	1908	119.5
1891	53.4	1900	100.0	1909	122.5
1892	55.7	1901	103.1	1910	141.4
1893	63.3	1902	103.8	1911	149.7
1894	63.3	1903	106.5	1912	153.2
1895	70.4	1904	109.5	1913	163.6

Source: Nove (1972, p. 12).

Table 1.3 National income in 1913: Russia and other selected countries

	National income (million rubles)	Population (millions)	National income per head (rubles)	Russia = 100
Russia	20,266	171	119	100
UK	20,869	36	580	487
USA	96,030	93	1,033	868
Germany	24,280	65	374	314
France	11,816	39	303	254
Italy	9,140	35	261	219
Spain	3,975	20	199	167

Source: Gatrell (1986, p. 32).

Table 1.4   Russian economic indicators, 1890–1914

	1890	1895	1900	1905	1910	1913
Pig iron production (million puds)	56.6	88.7	179.1	166.8	185.8	283
Coal extraction (million puds)	367.2	555.5	986.3	1,139.7	1,526.3	2,200
Use of raw cotton (million puds)	8.3	12.3	16.0	18.2	22.1	25
Railway net (thousand km)	30.6	37.0	53.2	61.1	66.6	70
Imports (thousand rubles)[a]	406.650	526.147	626.375	635.087	1,084.442	1,374.0
Exports (thousand rubles)[a]	692,240	689,082	716,217	1,077.325	1,449.084	1,520.1

Note: [a] Visibles.
Source: Munting (1982, p. 32).

themselves. On the other, Russians engaged in the economic exploitation of various southern and eastern parts of the empire as well as of
contiguous territories. These patterns are reflected in the country's economic geography, with industry tending to cluster around Moscow or
close to the western frontiers (figure 1.13) (Spechler, 1980).

Despite such developments, Russia's industrial progress by 1914 was
still only partial. As Lenin wrote in 1899: 'The process of transformation
must, by the very nature of capitalism, take place in the midst of much
that is uneven and disproportionate' (quoted in Gattrell, 1986, p. 231).
Even in the relatively few centres where the most modern industry was
concentrated, life still had many traditional features, such as seasonal
flows of population between rural and urban areas. A good deal of
manufacturing was still to be found in the countryside, and agriculture
was often extremely backward. In both town and countryside, modernization was accompanied by social distress, reaching boiling point during
the revolutionary years 1905–6. Such social currents proved an obvious
menace to the political future of the tsarist regime.

Given these circumstances, it is not surprising that the experience of
industrialization provoked sharply contrasting reactions among social
commentators. Peter Gattrell has divided these people into three groups
(Gattrell, 1986, pp. xiv–xvi). The first group, liberals like Sergey Witte,
welcomed industrialization as the means whereby Russia would be able
to modernize and assert its status as a great European power alongside
Germany, Britain and other states. By contrast, the populists feared
capitalist industrialization as inappropriate to Russian circumstances
and potentially harmful to the peasants, who formed the great majority
of the population. Finally, the Marxists, such as Lenin, regarded capitalist industrialization as an inevitable, historical process which would

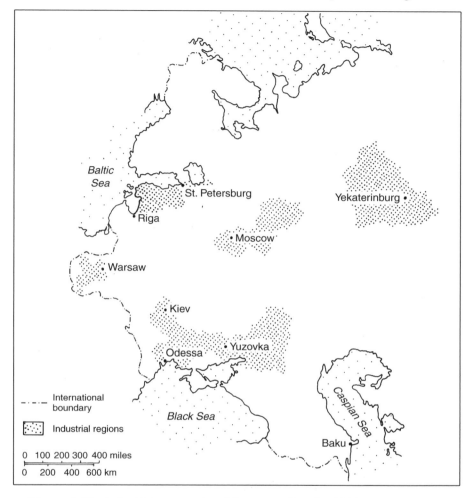

Figure 1.13   Industrial geography of European Russia, 1913.

hasten the day when a socialist revolution could be expected. Lenin and other Marxists debated with populists and liberals about how far capitalism had progressed in Russia in the years before 1914 and whether it would ultimately succeed in transforming Russian society (Mendel, 1961; Walicki, 1969). Such uncertainties are reflected in the present-day literature, some historians arguing that Russia would have followed the Western route towards modernization had it not been for the 1917 Revolution, others that it was becoming more and more like a Third World nation and yet others that Russia needs to be viewed not as an individual case of development but in the context of its role in the capitalist world economy (see Pallot and Shaw, 1990, p. 265). What concerns us here is to note that Russian experience before 1917 was the product of a combination of forces operating at the world scale (the development of capitalism, of world markets and of the interplay of international politics) and of circumstances and policies peculiar to Russia itself. By 1917 that combination had produced a very partial pattern of industrialization, a partial penetration by the forces of capitalism. It was against that background that the dramatic events of the Russian Revolution unfolded.

Two

The Soviet Heritage

The year 1917 witnessed a major turning point in Russian (or, as it was now to become, Soviet) development. From that time, and for most intents and purposes until the late 1980s, the Soviet Union turned its back on the outside world, certainly the capitalist world, in the attempt to build its own, distinctive socialist system. The challenge posed so long before by the Westernizers, advocating that Russia modernize itself and copy the advanced Western countries, was not forgotten. But Russia was now to do this in its own distinctive way, to follow its own destiny, in a fashion reminiscent of the Slavophils. In some ways the transfer of the country's capital from Petrograd (as St Petersburg was called from 1914 to 1924) back to the ancient city of Moscow, an event which took place in March 1918, symbolized this new orientation. In place of the semi-modernized regime bequeathed by Peter the Great which ruled the country from St Petersburg, his 'window on the West', there now arose a new state, established and centred in Moscow by V. I. Lenin and soon to acquire its own characteristic form under the harsh direction of Joseph Stalin. Ironically, this Stalinist state derived at least some of its symbols and outlook from the pre-revolutionary heritage, not the least significant symbol being the fact that it was now ensconced in Moscow's Kremlin, the historic seat of the medieval tsars. Yet it would be a mistake to dwell too much on the continuities from the past and to minimize the revolutionary character of Lenin's contribution and particularly of that of Stalin. Although Lenin was the founder of the Soviet state, Stalin was its chief architect, and his handiwork long outlived him. Part of that handiwork included a withdrawal from the world outside. The sweeping changes which began after 1985 consisted in the attempt to demolish a part of what Stalin had built and to reopen the window he so resolutely closed.

Revolution: the Establishment of the Soviet State

The Marxists were arguably the most significant of the many revolutionary groups which arose in late nineteenth-century Russia in response to the troubles of tsarism. The Russian Marxists had been attracted by the seemingly scientific character of Karl Marx's teaching and in particular by his insistence that the progress of capitalism would eventually result in its own downfall, ushering in the socialist era. Russia, with its immense social problems and patchy economic record, proved fertile ground for such an optimistic creed. Yet there were some real intellectual challenges for the Russians in Marxism. In particular, since Marx had based his teaching on the experience of western Europe, where industrialism was already well advanced, there was a serious doubt about how far his ideas were applicable to a country like Russia, where the overwhelming majority of the population were still rural peasants. The Russian Marxists took comfort from Russia's dizzy industrial progress during the three decades or so preceding the First World War. But, as we have seen, that progress was still not far reaching enough totally to transform Russian society. Lenin, who was leader of the Bolshevik fraction of the Russian Marxists, attempted to mould his Marxism to Russian conditions by arguing for a political alliance between the urban workers (Marx's revolutionary proletariat) and poor peasants, and by attempting to organize his party as a group of revolutionary activists. During the critical days of 1917, he was prepared to advocate an immediate seizure of power by the Bolsheviks, despite the belief of many Marxists that backward Russia was not yet ready for socialism. In the event, Lenin's opportunism ensured victory for his particular version of revolution.

The dramatic events leading up to the 1917 Revolution need not detain us. Suffice it to say that, undermined by military defeat in the First World War and the ensuing social turmoil, the tsarist regime fell in March of that year. It was immediately replaced by a shaky, middle-class Provisional Government, and by a system of semi-official councils or soviets. The latter functioned in the name of the workers, soldiers and peasants. The war continued and the problems of the Provisional Government mounted steadily. Then, in November (or October by the old calendar), it too was overthrown by the Bolsheviks who seized power in the name of the soviets. Within a few months, the Bolsheviks had dispensed with the Constituent Assembly, which had been elected to establish the basis for parliamentary government. They had also disbanded most of the opposition groups and established themselves as virtually the only legitimate political party. They now ruled Russia in the name of the people's revolution, and they meant to transform it along socialist lines.

The American historian Sheila Fitzpatrick has argued that the term 'Russian Revolution' ought properly to denote not merely a set of events occurring during 1917 but an entire process of political, economic and social change lasting from 1917 until 1932 (Fitzpatrick, 1982). During this period the essential structures of the Russian Empire were swept away and replaced by a totally new kind of state. A geographer inspecting the map of the USSR as it was in 1932, and comparing it with one of the Russian Empire in 1914, would see that the two occupied virtually the same territory. This fact suggests much in the way of ethnic, cultural and social continuity, as noted above. But it would be hard to exaggerate the profundity of the changes that swept across the country in the years following 1917.

It is important to understand that these changes were not merely the product of Marxist ideology. Of course, the Bolsheviks were Marxists and ideology played an important role in their thinking. Thus they viewed Russia as a society divided into classes, and the essence of a socialist revolution as far as they were concerned was the removal of the control of wealth and political power from the hands of the exploiting class of aristocrats, capitalists and their supporters, and its transfer to the vast majority of the population, the workers and peasants. But there were major problems in deciding how the people were to exercise their new-found powers, especially since many of them clearly had little under-standing of their own class interests as the Marxists defined them. Marxists tended to talk of a 'dictatorship of the proletariat', under which the working class would dominate its former masters, the bourgeoisie. Many expected that this dictatorship would be exercised by the soviets. In practice, it was quickly assumed by the Bolsheviks.

As good Marxists, the Bolsheviks believed not only in theory, but also in adapting their beliefs to Russian conditions. Lenin's insistence before 1917 on organizing the Bolshevik Party as a group of revolutionary activists tended to encourage the idea, once it seized power, of the party being the natural leader of the working class. It may also have fostered an unwillingness to tolerate those who dissented from the party line. Many scholars have seen in such tendencies a continuation of the old Russian traditions of autocracy and central control. Centralization may thus have accorded both with Russian political culture and with the natural inclinations of the Bolsheviks. But it was also soon to be fostered by events in the real world.

The Bolsheviks seized power in a situation of social turmoil. The First World War and the fall of the tsarist regime had helped to undermine belief in the legitimacy of the social order, and society soon began to descend into chaos. The seizure of power in Petrograd was followed by similar events across Russia, sometimes accompanied by violence. Power often fell into the hands of the local soviets, but the Bolsheviks frequently lacked influence at the local level. The struggle to establish social order was a difficult one, and in these circumstances the temptation to punish or repress one's opponents was not easily resisted. The Bolshevik attitude was also moulded by other considerations. They now claimed to rule a country, the majority of whose population were peasants either indifferent or deeply hostile to Bolshevik ideas (despite Lenin's hopes of an alliance between urban workers and poor peasants, Marxism never had much appeal in the countryside). This may have contributed to a siege mentality among the Bolsheviks. The international outlook was also distinctly unpromising. The Western powers were alarmed by the establishment of an avowedly socialist regime in Russia, and particularly displeased by its insistence on making a separate peace with the Germans in March 1918, thus abandoning Russia's former allies. The simmering social unrest and the numerous conspiracies and intrigues quickly exploded into civil war, abetted by the intervention of foreign troops. The three years of civil war (1918–21) did much to endow the Bolshevik regime with an authoritarian and rigorous outlook.

The civil war witnessed the dismemberment of the old Russian state and its recombination into the new Soviet Union. Throughout the period, the Bolsheviks held on to the old core of Russia, Moscow and its surrounding territories. They were opposed in varying degree and at varying times by different combinations of 'White' (i.e. anti-Bolshevik), non-Russian nationalist and foreign forces which held the peripheries. Eventually, after a series of Bolshevik victories, the Whites were abandoned by their foreign supporters and left to their fate. The nationalists were either defeated or successfully wooed by the Bolsheviks.

One of the most striking Bolshevik victories of the period was their success in containing most of the numerous secessionist movements which threatened to dismember the old tsarist empire. The Bolsheviks' declared belief in equal rights for all nations, and their avowed opposition to colonialism and imperialism, were tempered by a class-based approach to the issue of self-determination for national groups. The Bolsheviks were internationalists rather than nationalists, and believed that the workers' revolution should lead to the breaking down of national barriers and an end to ethnic rivalries. Where a bid for independence was made by a well organized bourgeois movement which effectively made its will felt, as in Poland and Finland, the Bolshevik government was forced to accept it. Polish independence had already been recognized by the Provisional Government and was officially accepted by the Bolsheviks after the October Revolution. Finnish independence followed in December 1917, but not without some subsequent oscillations in the Bolshevik attitude. The same uncertainties attended recognition of Estonian, Latvian and Lithuanian independence, which occurred through a series of treaties signed in 1920. The problem lay in the Bolsheviks' class-based approach to the national question. Where bourgeois demands for national independence, and in particular for national power, were opposed by a workers' movement, the natural inclination of the Bolsheviks was to support the latter. Even more significant was the fact that the Bolsheviks were determined to protect the revolution from its enemies. Across the former Russian Empire, national movements became bound up with the complexities of civil war and the fighting between bourgeois, socialist and sometimes semi-feudal interests. The Bolshevik triumph in the civil war meant victory over bourgeois nationalists and others who opposed the revolution. This was usually achieved in alliance with workers' and other 'progressive' forces, sometimes only a small minority operating at local level. In time, a series of workers' republics came into being which signed treaties among themselves and with the Russian Republic (now to be known as the RSFSR or Russian Federation). The creation of the Union of Soviet Socialist Republics (USSR or Soviet Union) in December 1922 was the logical culmination of this process of reunification. Initially, this consisted of four nominally equal Union republics: the Russian Federation (RSFSR), Ukraine, Belarus' (or Belorussia) and the Transcaucasian Republic.

By the end of the civil war, the greater part of the territory which had formed the old Russian Empire had been recombined into the USSR (figure 2.1). Unlike the old empire, the USSR was a federation of supposedly equal republics. However, the federal structure of the new state was in essence counteracted by Bolshevik centralization. Gradually, the Bolsheviks, ruling from Moscow, made their power effective across the entire territory and in the different republics. The Bolshevik Party (henceforth to be known as the Communist Party) was centralized under the control of Lenin and his associates, and the various workers' organizations, like the soviets and the trade unions, were gradually reduced to subservience. The governments of the different republics were subordinated to Moscow. New instruments, like the formidable Red Army, organized by Leon Trotsky, and the Cheka or secret police, also served to consolidate communist control. Even more important in the long run was the process of economic centralization. It is to this question that we must now turn.

Economic Transformation

No more than in the political sphere was Bolshevik policy for the economy purely determined by Marxist ideology. Marxist ideology did prescribe some economic measures, above all that the sources of wealth

Figure 2.1   The USSR in about 1923.

(the 'means of production') be taken out of the hands of the exploiting classes and assumed by the proletariat. But Marxism laid down no blueprint about precisely how this was to be done, or about policies to ensure future economic growth. Unlike some of their revolutionary rivals, the Marxists were modernizers, believing in the need to industrialize and to benefit from the experience of the advanced capitalist powers. Yet they rejected the capitalist road taken by the West and by the pre-revolutionary tsarist regime. The West, as one writer has argued, was both a model and an anti-model for the Bolsheviks. 'Like Peter the Great and other modernizers, the Bolsheviks hoped to take Western technology without Western values' (Sakwa, 1989, p. 33). The route finally chosen was more the product of short-term political considerations than of long-term economic or Marxist-theoretical ones.

Early policies were rather cautious as the Bolsheviks felt their way in the new situation (Davies, 1994). Land was nationalized immediately and private estates were abolished, thus securing at least the neutrality of the peasants. Banks were nationalized and a gradual nationalization of some other activities and enterprises also took place (not always on the initiative of the government). However, this gradualist policy was rudely interrupted by the outbreak of the civil war in the middle of 1918. The grave difficulties this entailed spurred the regime into a wholesale nationalization of industrial activity and the forced requisitioning of food from the peasants. Private trading by the latter was forbidden. The effect of this drastic policy was to undermine the alliance between the urban workers and poor peasants which Lenin had proclaimed as an essential of the revolution. The policy, which was justified by many in terms of class warfare, came to be known as 'War Communism'.

The civil war was won only at enormous cost to the economy. In 1921, therefore, Lenin introduced a radical policy change, known as the New Economic Policy (NEP). This was designed to restore production to pre-war levels. Under NEP, private trading was once more permitted, the aim being to restore the alliance (*smychka*) between town and country. Some industry was denationalized, especially small-scale industry and handicrafts. But the state retained in its hands the 'commanding heights' of the economy, including major industry, the banks and foreign trade.

Economic recovery followed quickly under NEP and by about 1926–7 pre-war production levels had been restored. The problem now was how to move forward. Lenin, the initiator of the October Revolution, had died at the beginning of 1924 without having made it clear whether he regarded NEP as a short-term expedient or a long-term policy due to last for decades. His death opened the door to political succession struggles among the other party leaders, and these political struggles were fuelled by disagreements about economic policy. The problem was that Marxism allowed for different opinions about the correct policies for economic development and, since the Soviet Union was the first country to try to develop within a Marxist framework, there was no previous exemplar to follow. Two main protagonists emerged. The first, known as the rightists and led by N. I. Bukharin, argued for a continuation of NEP-type policies as the best way to ensure economic growth within a socialist framework in what was still basically a peasant society. The second, known as the leftists and including E. A. Preobrazhenskiy (and for a time Leon Trotsky), asserted that NEP-type policies encouraged capitalism in the countryside, thus endangering the revolution, and were unlikely to achieve the rate of economic growth required. They argued for a policy of rapid industrialization and stringent controls on peasant agriculture. Many felt that such measures would barely succeed unless the USSR were aided by a friendly foreign country with a more advanced economy (hence the need to spread the revolution abroad).

The USSR's international position had an important bearing on these arguments. Lenin had seized power in 1917 in the hope that a socialist revolution in Russia might prove a catalyst to revolution abroad (years before, Marx had written of Russia as a possible weak link in the capitalist chain) and that Russia might thus be helped to build socialism by more advanced countries in the West. This hope was dashed by the early 1920s, as it became clear that no such social upheaval could be expected in the West, at least in the near future. In the meantime, the capitalist powers had been antagonized by the Bolshevik repudiation of the foreign debts of the previous regimes, their nationalization without compensation of foreign property, the peace treaty with Germany signed in March 1918 and perhaps above all by Soviet support for foreign revolutionary movements. Lenin's subsequent attempt to sell 'concessions' in Soviet resources to Western businesses, and to revive foreign trade, met with little success. The USSR was almost totally isolated in the 1920s and had little hope of deriving much help for development from abroad. It was this situation which was used by Joseph Stalin, party general secretary from 1922, to outmanoeuvre his opponents and establish himself in a position of supremacy within the party. Since Trotsky and other leftists were looking to international socialism to help the revolution in the USSR, Stalin was able to accuse them of lacking patriotism and faith in the Soviet people's ability to build socialism without foreign aid. He thus advocated what came to be known as 'socialism in one country'. Later he turned on Bukharin and the rightists, whom he accused of condemning the country to years of 'snail's pace' economic growth under NEP-type policies.

In the end, Stalin opted for a strategy of more rapid economic growth than even the leftists had advocated. This entailed the termination of the mixed economy as it had operated during the NEP period, the completion of the nationalization process to embrace even private handicrafts and retail trade, and the introduction of central planning, or what has been termed the 'command economy'. In agriculture, the pressures produced by the policies of forced industrialization inevitably provoked opposition, and the requisitioning of food was resumed. The government then resorted to the compulsory collectivization of peasant holdings, accompanied by full-scale class war in the countryside. The results of this catastrophe are still to be felt today. The many economic changes, constituting a veritable 'revolution from above', began to become apparent from about 1926, but especially with the introduction of the first five year plan in 1928. They were largely completed by the early 1930s.

Stalin's policies were extraordinarily successful in transforming the Soviet economy and dramatically increasing industrial production (table 2.1), albeit at considerable human cost, including famine and the widespread use of forced labour. The overall economic policy, which has been described by Senghaas, in a rather inelegant phrase, as 'dissociative state-socialist development' (Senghaas, 1985), had three central characteristics. First, it was 'state-socialist' because it was development through state control and planning rather than through private capitalism. Second, it involved a high degree of self-sufficiency, importing only the minimum necessary in the way of technology and temporary skilled labour, and paying for them with exports of foodstuffs, raw materials and other goods. This policy had the advantage of sheltering the Soviet economy from the uncertainties of the world market and the Soviet state from the politically suspect machinations of the capitalist powers. The third characteristic was that it was a policy of unbalanced economic growth, emphasizing some sectors (especially producer goods) and neglecting others (agriculture, transport, consumer goods). It is thus possible to speak of a Stalinist model of economic development, but the planned nature of that development should not be exaggerated. There was much chaos and improvization in the planning process (what has been called 'wilful planning' (Grossman cited in Sakwa, 1989, p. 44)) and the accompanying social upheavals were only partly understood and controlled.

The geographical consequences of Stalinist economic development were far reaching as the policy of unbalanced growth was reflected in regional patterns. Compulsory collectivization was a clear breach of Lenin's principle of forging an alliance between the urban workers and poor peasants. Far from bridging rural–urban differences, a Marxist aim ever since the 1848 *Communist Manifesto*, this action drove the two further apart. Spontaneous movements of people from the countryside alarmed the government and brought chaos to the cities. Urbanization reached explosive dimensions. No less significant were the changes which occurred at regional level (figure 2.2). Because Stalin was determined to industrialize as quickly as possible, maximum use had to be made of already-existing industry and infrastructure inherited from the tsarist regime. Thus those regions which were already industrially developed before 1917, such as the Northwest, the Central Industrial Region around Moscow, the southern Ukraine and the Urals, continued to develop during this period. However, partly for strategic reasons, the Stalin era also witnessed considerable spatial spread in the industrializa-

Table 2.1   Selected indicators for the first and second five year plans (1928–37)

	1928	1932/3 (plan)	1932 (actual)	1937 (plan)	1937 (actual)
National income[a]	24.4	49.7	45.5	100.2	96.3
Gross industrial production[a]	18.3	43.2	43.2	92.7	95.5
Gross agricultural production[a]	14.5	22.6	13.1	26.2	20.1
Workers in state employment (millions)	11.3	15.8	22.9	28.9	27.0
Electricity (millions kWh)	5,050	22,000	13,540	38,000	36,200
Coal[b]	35.4	75	64.4	152.5	128.0
Oil[b]	11.7	21.7	21.4	46.8[c]	28.5[c]
Steel[b,d]	4.0	10.4	5.9	17.0	17.7
Pig iron[b]	3.3	10.0	6.2	16.0	14.5
Machine tools (thousands)	2.1		19.7	40.0	45.5
Tractors (thousands)	1.8		50.6	166.7	66.5

Notes: [a] Billion rubles at 1926–7 prices; [b] million tonnes; [c] includes natural gas; [d] ingots and castings, excluding rolled steel.
Source: Munting (1982, p. 93).

tion process and the increasing exploitation of remote territories for their resources. Even so, such broad patterns hide the many spatial discrepancies which Stalin's narrow economic emphasis helped to foster.

Stalinism and After

Stalin transformed the Soviet economy, but in doing so he also transformed the Soviet state. Historians have disagreed about the extent to which the events accompanying Stalin's rise were an inevitable and logical outcome of Lenin's earlier policies rather than being caused by such contingent factors as Stalin's own personality. What concerns us is to note the most prominent features of 'Stalinism': his personal dictatorship over the country and even over the Communist Party, the regular use of purges and terror, the growth in importance of the secret police (the NKVD, ancestor of the KGB), the elimination of autonomous organizations, ruthless repression of any sign of nationalism and the attempt to harness almost every facet of society to political ends. In the

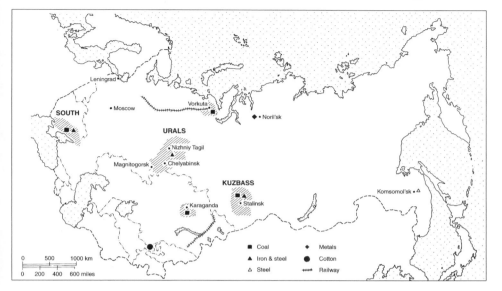

Figure 2.2   Major regional economic developments, 1920s to 1950s.

post-war period, Western political scientists coined a new phrase to describe the type of political system Stalin had created: 'totalitarianism' (Arendt, 1958; Lane, 1982, pp. 146–51). It was, to quote one commentator, a 'distorted' form of modernization which emphasized some aspects of what has been termed the modernization process (for example, industry and military prowess) while repressing others (freedom of speech, criteria for rational decision-making, creativity) – an 'anti-social socialism' (Sakwa, 1989, p. 59).

An additional feature of the Stalin period was the transformation it brought to the USSR's international position. At the end of the 1930s, through various manoeuvrings which included the notorious Molotov–Ribbentrop pact with Adolf Hitler in August 1939, Stalin was able to regain a number of territories in the west which had been lost after the revolution (most notably, the three Baltic states of Estonia, Latvia and Lithuania, annexed by the USSR in 1940) (figure 2.3). But this did not save the Soviet Union from the catastrophe of the German invasion in June 1941, and near-defeat in 1941–3. By the end of the Second World War in 1945, the country had been devastated by enormous human and material losses. Yet its armies were in occupation of Berlin and of much

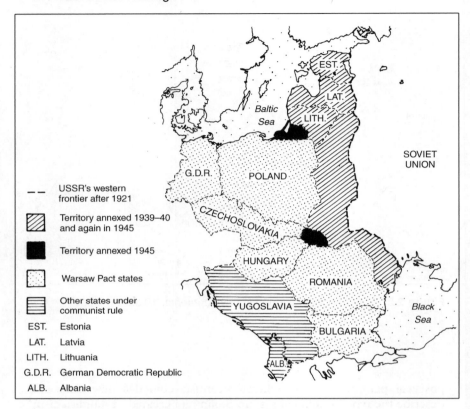

Figure 2.3   USSR and eastern Europe after 1945.

of central and eastern Europe. In succeeding years a series of communist regimes was established in these 'liberated' territories. The countries concerned were closely allied with the USSR in the Warsaw Military Pact (1955) and in a new economic organization, the Council for Mutual Economic Assistance (CMEA or COMECON) (figure 2.3). Soviet prestige and influence reached new heights as communism gained ascendancy in parts of Asia, especially China, and further afield. Predictably, all this revived mutual suspicion and hostility between the USSR and the Western capitalist powers, particularly the United States, whose own international prominence had been similarly promoted by the war. Thus began the era of the Cold War. The period witnessed an arms race as both power blocs stockpiled nuclear weapons and other armaments in case a real war should break out once again.

Table 2.2   Average annual growth rates in the post-Stalin period

	1951–5	1955–65	1966–70	1971–75	1976–80	1981–3	1984–7
GNP	5.5	5.4	5.2	3.7	2.7	2.3	1.6
Industry	10.2	7.5	6.3	5.9	3.4	1.5	2.1
Agriculture	3.5	3.5	3.5	−2.3	0.3	4.2	0.8
Services	1.9	4.0	4.2	3.4	2.8	2.1	
Consumption	4.9	4.7	5.3	3.6	2.6	1.7	2.4
Investment	12.4	9.1	6.0	5.4	4.3	4.2	3.0

Source: Gregory and Stuart (1994, pp. 128, 133).

Stalin's death in 1953 removed from the scene the chief architect of the Soviet state (Keep, 1995). His successors, Nikita Khrushchev (leader from the mid-1950s to 1964) and Leonid Brezhnev (leader from 1964 to 1982) did introduce some domestic changes. On the economic front, they attempted to move away from Stalin's narrow emphasis by placing more weight on the need for industrial modernization, agriculture and the consumer sphere, albeit with relatively limited success. There were also important political changes, such as the reduced significance of the personal dictatorship and the 'personality cult' which had so characterized the Stalin years (in the USSR Stalin himself now became one of history's unmentionables). In addition, the powers of the secret police were somewhat curbed, the tendency to resort to arbitrary arrest and imprisonment was reduced and there was slightly greater opportunity for debate and political participation (there were fluctuations between more or less 'liberal' periods, just as in foreign affairs periods of détente alternated with more confrontational interludes). But the essentials of the Stalinist state remained, viz the one-party state, the centrally planned economy, collectivized agriculture and all kinds of restrictions on civic freedoms, democracy and travel. To the extent that Stalin was criticized, it was for his personal errors rather than for the basic institutions which he had bequeathed to his successors. Yet these institutions 'were endowed with a legacy of inflexibility' (Sakwa, 1989, p. 59) which made them increasingly unsuitable for the complex society that the Soviet Union was now becoming. One of the symptoms of that unsuitability was the slowing down of economic growth rates, which became ever more noticeable as time went by (tables 2.2, 2.3).

In the international arena, the Khrushchev and Brezhnev years were marked by the USSR's determination to be recognized as one of the world's two superpowers, the leader of world socialism and of an open or covert challenge to world capitalist supremacy. It was an extremely ambitious goal for a country which had wrestled for so long with the problem of relative backwardness. The question was whether Marxism-

Table 2.3 Alternative estimates of average annual economic growth rates, 1951–85

	Official Soviet	Selyunin/Khanin	CIA	Steinberg
1951–60	10.3	7.2	5.1	
1961–65	6.5	4.4	4.8	
1966–70	7.8	4.1	5.0	4.8
1971–75	5.7	3.2	3.1	2.1
1976–80	4.3	1.0	2.2	1.6
1981–85	3.6	0.6	1.8	1.0

Source: Gregory and Stuart (1994, p. 236).

Leninism had at last provided the solution to this historic problem. The answer was to be a negative one. Although Soviet influence was now to be felt directly or indirectly on every continent, and although by the 1970s Soviet naval forces were sailing in every ocean, the USSR succeeded neither in facing down the capitalist powers through victory in the arms race, nor in sustaining the world communist movement beyond a limited point. Indeed, it found it increasingly difficult to maintain its leadership in the international communist camp, as China's gradual adoption of an anti-Soviet stance illustrated only too well. Geopolitically, the United States and its allies largely succeeded through a policy of containment in confining most Soviet activities within a zone fringing Mackinder's Eurasian heartland (Taylor, 1993, pp. 92–8). As the 1980s advanced, and a more assertive American president seemed determined to regain initiatives lost by the USA in the previous decade, more and more Soviet citizens began to ask whether the USSR's now ailing economy could really sustain the world role it aspired to.

The day of reckoning might in fact have come sooner than it did, had it not been for two factors. One was the USA's partial loss of self-confidence following its disastrous defeat in the Vietnam War in the early 1970s. The other was the Soviet Union's success, as a departure from its traditional policy of economic autarky, in developing its international trade. Especially after 1970, it was able to use its growing export of hydrocarbons to pay for needed imports of technology, food and consumer goods. This allowed it not only to overcome bottlenecks in its own industrial development and agricultural performance, but also to promote long-promised improvements in living standards. Trade also permitted the USSR to cement its relations with the other members of COMECON and to further its influence in the Third World.

By the 1980s the limitations to the system which Stalin had created were becoming only too apparent. The command economy was consistently failing to perform in the expected way and to produce the wealth needed to fulfil the ambitious goals adopted by the party leaders. Some of the reasons for this will be explored in chapter 4. On the political level, the ageing party leadership was facing increasing difficulty in securing the loyalty and directing the aspirations of the citizenry. In the end, in March 1985, a relatively young party leader, Mikhail Gorbachev, was appointed with a brief to reform the system. His attempts to do so, dramatic as they seemed at the time, were soon overshadowed by the spectacular events which signalled his failure. By the end of the 1980s the entire Soviet bloc of communist states had fallen away, and the USSR itself was in an advanced state of dissolution. Finally, in 1991, the 15 constituent republics of the USSR achieved independence and the USSR itself passed into history. The heir to most of its land, resources and people was the Russian Federation.

General Reflections on Soviet Communism

A feature of a good deal of the geographical literature which surveyed the development of the world economy in the 1980s and early 1990s was the tendency either to ignore the Soviet Union and its allies entirely, or to assume that they were simply 'capitalist' states whose geographies required no special explanation (see Lloyd and Dicken, 1977, pp. 3–4; Knox and Agnew, 1989). What we wish to do in this conclusion to our discussion of Russia's Soviet heritage is to raise some questions challenging the soundness of this assumption. This is important because later in this book we shall be arguing that the current economic and social transition in Russia has been particularly difficult precisely because of the country's idiosyncratic history. The final part of this section will consider whether some of the geographical literature which explores the recent history of world capitalism also has something to teach us about the reasons for the decline and fall of the Soviet system.

Geographers who have written as if the Soviet Union were 'capitalist' may have been influenced by the many Western Marxists and others who have asserted that the Russian Revolution was not a genuinely socialist revolution, and that it merely replaced one type of capitalism by another. This argument would be based on the view that backward Russia was not ready for socialism in 1917, that Lenin had acted prematurely and that his attempt to initiate a socialist experiment was bound to give rise to highly distorted results. One version of this argument is the assertion that the Soviet system was an example of 'state capitalism'. Proponents of this view have often pointed out that the Soviet economy was just as exploitative in its own way as Western-style capitalism, and that it was run by a permanent class of privileged politicians and bureaucrats who were the equivalent of Western-type capitalists. Moreover, socially and spatially Soviet society was just as unequal as Western capitalist society, and those inequalities gave little indication of diminishing through time (see, for example, Knox and Agnew, 1989, p. 160). One of the best-known exponents of this kind of position is Immanuel Wallerstein (1979), who maintains that, since the Soviet Union formed part of the capitalist world economy (on his own, rather idiosyncratic, definition of 'world economy'), it was by definition capitalist. Not until the downfall of the capitalist system, an event which Wallerstein feels may not occur in the near future, can true socialism be hoped for.

The problem with the argument about 'state capitalism' is that it tends to use the proposition that the Soviet system was unjust and exploitative (which is why many people have refused to describe it as socialist) as the logical basis for asserting its capitalist character. The difficulty is that this minimizes the sweeping changes which occurred in Soviet society after 1917. Socially, these changes entailed the destruction of the old aristocratic and bourgeois classes and their replacement as the social elite of the Soviet system by a whole new class of bureaucrats, technocrats, political functionaries, scientists and skilled workers largely recruited from the industrial proletariat and the peasants. As time went by, social mobility declined and society became increasingly stratified (Lane, 1982). But it was impossible (at least as far as the law was concerned) for even the most privileged member of society to own large amounts of private capital or to inherit wealth in quite the same way that it is possible in the West. There was therefore no 'capitalist exploitation' of labour on the Western model (although other forms of exploitation existed), and the many social inequalities which did exist were based upon different processes.

There thus appears to be a need for a theory to describe the special kind of society which arose in the USSR after 1917, a society which probably took its most characteristic features from the attempt to impose a version of industrial socialism on the weakly developed capitalist and peasant foundations of the pre-existing system (Lewin, 1988). There is something to be said for the view that this type of society (albeit now widely judged as an historical failure) should be called 'state socialism'. This is not because such a society was juster and more 'progressive' than Western capitalism, and hence genuinely socialist. In fact the opposite seems certainly the case. It was rather that this society was endowed with certain fundamental 'socialist' features, such as a nationalized economy.

As an industrial society, it had numerous points of similarity to Western capitalism. But the differences were also quite fundamental and seem to require distinctive theoretical treatment (Ellman and Kontorovich, 1992).

This is not the place to provide such a theoretical discussion (for some attempts to do so, see Cliff, 1964; Davis and Scase, 1985; Thrift and Forbes, 1986, pp. 6–27). The spatial implications of the distinctive Soviet mode of development will be examined later in this book. Here we wish to pose the question why this distinctive society failed to endure, and to do so in the context provided by the literature on the recent evolution of the world economy. This will provide some theoretical background for the more detailed discussion of Soviet failure to be found in chapters 3 and 4.

To consider the Soviet system in the light of the evolution of Western capitalism might seem a strange thing to do in view of what has been written above, emphasizing the differences between the two systems. However, the fact remains that, for all the differences, Soviet society was an industrial society and, down to the 1960s at least, it shared many features with industrial capitalism. Thus there was a similar accent on extractive and manufacturing industry, on large-scale industrial plants, on industrial corporations controlled from a central point or points, on state regulation of the economy, on regional industrial specialization, on big industrial cities, on bureaucratic forms of control, on state controlled welfare provision and on much else. It has been argued by numerous scholars that these are all features which typified the 'Fordist' era of industrial capitalism (Sayer, 1989). It is argued that Fordism was characterized by a reliance upon mass production, standardized products and markets, economies of scale, scientific management ('Taylorism') and big corporations. All this, of course, was modelled on the automobile production methods pioneered by Henry Ford at the beginning of the twentieth century. Henry Ford's methods greatly interested the Soviets. As Murray wrote: 'Soviet-type planning is an apogee of Fordism. Lenin embraced Taylor and the stopwatch. Soviet industrialization was centred on the construction of giant plants, the majority of them based on Western mass-production technology' (Murray, 1988, p. 9).

Down to the 1960s, some scholars claimed to detect a growing convergence between Western and Soviet-type industrial societies as Fordist methods continued to influence both (Galbraith, 1969). After that time, however, the capitalist economies seemed to undergo some significant changes. Thus Lash and Urry claimed that since the 1960s the advanced capitalist countries have been moving from the stage of 'organized capitalism' to 'disorganized capitalism' (Lash and Urry,

1987). They list the fundamental features of the latter stage as follows: a decline in the relative importance of employment in extractive and manufacturing industry, especially by comparison with that in services, a reduction in the average size of industrial plants, with more accent upon labour-saving investment and more flexible employment processes, a tendency for industrial firms to 'hive off' many needed services and activities to other organizations, the reduced effectiveness of state regulation of the economy, challenges to the centralized welfare state, the reduced significance of regional economies, the reduced importance of big industrial cities and an increase in cultural fragmentation and pluralism.

The processes promoting these developments are complex and interrelated (see Bennett and Estall, 1991). Among those which have been cited as especially significant are the growing interdependence of capitalist countries within the global economy, the activities of transnational corporations, the flexibility offered by new technologies and the demands of an increasingly sophisticated 'service class'. Scholars have thus proclaimed the close of the Fordist era, based on mass production, and the arrival of a 'post-Fordist' or 'neo-Fordist' era of flexible production, with an accent on innovation, 'customization', design and quality (Massey, 1984, pp. 23–5; Amin, 1994).

The Soviet Union always regarded itself as the West's competitor, both politically and economically. The success of the Soviet system, and its ability to influence events in the rest of the world, seemed to depend upon overtaking the West in the 'peaceful competition' of the Cold War era. This is why the sluggishness of the Soviet economy after about 1970 caused such concern to the leaders in the Kremlin. Thus the Western tendency towards more flexible production systems from the 1970s, spurred by the introduction of modern technologies, was bound to find some echo in the Soviet Union. If the USSR was to continue to compete with the West for world influence, if it was to keep up militarily and if it was going to be able to benefit from world market developments (an arena which, as we have seen, the USSR increasingly entered after 1970), then it was essential to adapt to the latest technological developments. And this in turn meant trying to introduce some of the flexibility in research, development and production methods which was becoming apparent in the West and, more particularly, in Japan and other parts of east Asia. The old Stalinist planning system, with its high degree of vertical integration and a minimum of horizontal linkages between industrial enterprises, seemed clearly out of place in a 'post-Fordist' era. Arguably, the traditional system was simply too rigid and bureaucratic to foster a complex, modern economy.

Table 2.4 Labour force structure of western and
eastern Europe, 1955–85 (percentage shares)

Region/sector	1955	1970	1985
Western Europe			
Agriculture	20	15	10
Manufacturing	42	42	33
Services	38	43	57
Eastern Europe			
Agriculture	47	29	20
Manufacturing	30	39	42
Services	23	32	38

Source: Treyvish et al. (1993, p. 161).

In the West, the transition to 'post-Fordism' was partly a response to
the increased affluence of the population and especially of the middle
class, or that section of it sometimes referred to as the 'service class'.
Greater affluence tends to lead to greater demand for a more varied
range of services and goods, based less on mass provision and more on
individual need and choice. Soviet society did not experience a decline in
extractive and manufacturing industrial employment relative to service
employment to anything like the same extent as that seen in the West
(table 2.4), and hence the 'service class' was much less significant.
Nevertheless, it grew considerably as society industrialized and urbanized and as standards of living rose. The growth of the consumer sphere,
which was particularly marked from the end of the Stalin period down to
the 1970s, was partly a reflection of the influence of Western values and
ways of life, and partly of the leadership's realization that material as well
as moral incentives are essential to economic progress. In this context it
was the attitudes and outlook of the professional, technical and executive
strata, known collectively as the 'intelligentsia', which were decisive. The
numbers in these groups grew considerably after the 1950s, since they
are vital to a modern industrial society (Lane, 1982, p. 117). The advent
of modern technologies could only increase their importance. As in the
West, as time went on these groups seemed likely to become more
differentiated in their tastes and aspirations, less satisfied with standardized goods and services, and more desirous of enjoying democratic
freedoms and greater autonomy both at the workplace and elsewhere.
None of this seemed reconcilable with the old, Stalinist system.

While both Khrushchev and Brezhnev tampered with the Stalinist
system, it was left to Gorbachev after 1985 to attempt more fundamental

reform. His ultimate failure can be blamed on the inability of the system bequeathed by Stalin to cope with the forces of change. A modern, technologically sophisticated society seemed to demand room for more autonomous, discerning and creative individuals and groups than could be catered for by a Stalinist system which had been invented for quite a different purpose and in an entirely different historical epoch.

There is in fact considerable evidence that the Stalinist system was becoming unstable even before 1985. Not only did technological advance engender a struggle between the bureaucrats running the party–state structure and the growing number of technocrats, but the authorities seemed to have only partial control over the outlook, lifestyle and aspirations of citizens, especially in the bigger cities (Lewin, 1988, p. 147). The 'disorganized capitalism' thesis suggests that Western society has gradually become more pluralistic and fragmented. A similar, if less obvious, process seems to have occurred in the USSR as attempts by the Communist Party to impose a unitary ideology and lifestyle were increasingly frustrated. As Lewin demonstrates, Soviet society became quite heterogeneous, with elements of a pre-industrial rural culture and an early industrial culture rubbing shoulders, especially in the big cities, with the scientific and cultural revolution of an advanced industrial society. This imposed enormous strains on the authoritarian, Stalinist structure. Just as in advanced capitalist societies, the need to modernize demanded massive readjustments. In the event, Gorbachev's attempts to procure those readjustments proved fatal to the system which Stalin had constructed. Quite how this occurred will be considered in the next two chapters.

Three

The Emerging Federation

Thus far much of this book has been concerned with Russia's long-term development, both in its manifestation as the Russian Empire before 1917 and in its twentieth-century manifestation as the USSR between 1917 and 1991. We now turn to look at the Russian Federation itself. The purpose of this chapter is to consider the initial establishment of the Russian Federation, or Russian Soviet Federal Socialist Republic (RSFSR), at the time of the Russian Revolution, its development before 1985 as part of the USSR and its rapid emergence as a distinctive political entity during the momentous years between 1985 and 1991. More extended attention will then be paid to the country's ethnic and political geography and to the problems which this has produced since the collapse of the USSR in 1991.

'Core and Metropolis': the Russian Federation before 1985

the RSFSR – is the central core and metropolis of the Soviet Union, rather than just one of its constituent republics.

Vadim Medish, 1980

The social turmoil and conflict which accompanied the 1917 Revolution and the ensuing civil war have been discussed in the previous chapter. Obviously, in those circumstances the political constitution and even the boundaries of the new society or societies which were coming into existence were matters of extreme uncertainty. What made things even more difficult was the rise of nationalism among some of the subject peoples of the former empire, and the ambivalent approach adopted by the Bolsheviks towards nationalities issues. As we have seen, the Bolsheviks were internationalists at heart who deprecated nationalism and who tended to think of the union of different peoples within large states as historically progressive. At the same time, for reasons of political expediency, it was necessary to make concessions to national feeling. Even as early as January 1918, in their Declaration of the Rights of the Toiling and Exploited People, the Bolsheviks had explicitly called the new Soviet state 'a federation of Soviet national republics' (Hosking, 1992, p. 98). The Russian Federation's first constitution, adopted in 1919, also acknowledged the federal principle. Although federalism was not part of the Bolshevik creed and was regarded as a temporary concession to nationalism, Lenin and other Bolshevik leaders came to the conclusion that it was a necessary stage in the gradual evolution of a united, international society of working people (Carrere d'Encausse, 1992, p. 114).

This was not a conclusion which was reached easily and without dispute. Towards the end of the civil war, the question arose about the desirable relationship between the RSFSR and the new workers' republics like Ukraine and Belarus', which had also been part of the Russian Empire but which had asserted their national identities and even known temporary independence during the civil war period. Stalin, who was Commissar for Nationalities, argued for their incorporation into Russia as autonomous regions. Lenin, however, believed that they should enjoy an equal status with Russia, at least in outward appearance, and that they should join Russia in a federation of Soviet republics. In the event Lenin's view prevailed and the USSR came into existence in December 1922. It seems that Stalin was most concerned with security matters and feared that too many concessions to nationalism might lead to the fragmentation of the Soviet state. Lenin, however, feared equally the spectre of Russian nationalism and the consequences of trampling upon the national sensibilities of non-Russians, sensibilities which were shared by many communists.

Whatever may have been the outward appearance of the new Soviet federation, however, it was in reality a highly centralized state, as noted in the previous chapter. In fact, the RSFSR was very well placed to dominate the other Union republics (which, together with Russia itself, numbered fifteen during the last 35 years of the USSR's existence). In the 1920s, before the creation of new republics in Central Asia and the reacquisition of lost territories in the west, the RSFSR held 90 per cent of the USSR's territory and 72 per cent of its population (figure 2.1). In addition, 72 per cent of the members of the Communist Party were Russians (Hosking, 1992, pp. 117–18). Russians dominated the Soviet state and moved happily between the republics, usually retaining their own language and culture as they did so. In fact, possibly a majority of Russians came to regard the whole of the USSR, and not just the

Russian Federation, as their homeland. They thus retained the imperial outlook of their pre-1917 forebears. Until late in the Soviet period, that outlook was reinforced by the tight control which Moscow exercised over the life and aspirations of the non-Russian republics.

Because the RSFSR was so big, it inevitably contained many nationalities other than Russians. In the early days, before the character of the future USSR had been determined, it was decided to grant some of the more significant of the minorities within Russia their own autonomous territories. The first of these was the Bashkir Autonomous Soviet Socialist Republic (ASSR), which came into existence in March 1919. This was followed in the following year by the Tatar ASSR. In the course of time, other ASSRs were organized, together with autonomous oblasts (regions) and autonomous (or national) okrugs (districts) for less numerous peoples. In accordance with the hierarchical principle which was now adopted for Soviet administrative regions, ASSRs had more (theoretical) rights than oblasts, and the oblasts more than okrugs (an autonomous okrug was usually administered by a neighbouring oblast centre). But the populations of all these autonomous territories came to enjoy special representation in the Soviet of Nationalities, one of the two houses of the Supreme Soviet or parliament. In addition, between 1926 and the late 1930s when they were abolished, a network of national soviets was established at local or village level to represent the interests of minorities without recognized territories of their own. In all these ways, then, the RSFSR was federalized, becoming a federation within the larger federation of the USSR. Needless to say, the amount of autonomy allowed to these lesser units within the RSFSR, and to their equivalents within some of the other Union republics, was very restricted indeed.

While insisting upon a high degree of political centralization, early Bolshevik policy was generally sympathetic towards the linguistic and cultural rights of nationalities, and much was done to promote education, health and cultural identity (Simon, 1991; Carrere d'Encausse, 1992). However, Bolshevik tolerance of cultural distinctiveness was limited by two factors. First, their fear of nationalism led them to pursue a policy of 'divide and rule' (Carrere d'Encausse, 1992, pp. 173–94; Zaslavsky, 1993, p. 35). Thus peoples who might wish to unite to express some broad cultural affinity, like the numerous peoples of Turkic or Mongol origin, were carefully subdivided between different autonomous territories and regions. Similarly, a great deal was done to frustrate linguistic unity by encouraging local dialects to become distinct languages. Second, the materialistic world view of the Bolsheviks and their resolute belief in the virtues of modernization made them generally unsympathetic to religious practices and traditional ways of life. Tribal peoples, like those living in Russia's far north, and Muslim groups were among those to suffer as a result of these attitudes. The Bolsheviks were thus prepared to manipulate their nationalities policies in favour of their broader goals of modernization and socialist unity. Even so, by granting territorial and cultural identity to peoples who had sometimes lacked any sense of distinctiveness previously, the Bolsheviks may have laid the foundations, albeit unwittingly, for a future upsurge of ethnic and national feeling.

In the threatening international climate of the 1930s, Stalin pursued a much harsher policy towards the nationalities. Industrialization and forced collectivization dealt a severe blow to many cultures, especially the more traditional ones, while numerous local leaders were accused of nationalism and other crimes, and disappeared in the purges. During the Second World War, entire ethnic groups, including the Volga Germans and several of the North Caucasian peoples, were accused of treachery and exiled to Siberia and Central Asia. Stalin also began to favour policies of Russification and to identify Soviet with Russian national identity. The Russians and their culture were increasingly identified as the first among equals. Stalin's successors, Khrushchev and Brezhnev, moderated many of his excesses but still pursued policies to encourage merger between the nationalities, often amounting to Russification. This was particularly the case within the Russian Federation. Here policies to promote Russian speaking and various restrictions on cultural expression were often more evident than in the other Union republics.

The historian E. H. Carr wrote that the Soviet Union was 'the RSFSR writ large' (quoted in Hosking, 1992, p. 118). During most of the Soviet period, in other words, Russians tended to identify with the USSR rather than the Russian Federation, and the Russians regarded themselves as the natural leaders of the country. The fact that the RSFSR lacked a number of the appurtenances of statehood common to the other Union republics, such as its own distinctive capital city, a republican branch of the Communist Party, many major government ministries and a national anthem, was of no importance. It merely signified the dominant position of the RSFSR: the USSR's instruments and emblems of statehood were for most purposes those of the RSFSR. In other words, the Russian Federation had virtually no separate existence. This was to change after 1985 in the most remarkable way.

The Resurrection of Russia: the Gorbachev Years (1985–1991)

When Mikhail Gorbachev assumed power as General Secretary of the Soviet Communist Party in March 1985, he faced the gigantic task of reforming what seemed to be an increasingly moribund system. Economic growth had been slowing since at least 1970, while there was a whole series of social, political and international problems in urgent need of attention. In his book *Perestroika*, published in 1987, Gorbachev spoke of a society 'ripe for change' and 'verging on crisis'. He went on to describe a host of problems which ranged from economic stagnation to a 'decay in public morals', accompanied by increasing cynicism and corruption (Gorbachev, 1987, pp. 17–25). What Gorbachev was describing was a society in real trouble, a country whose failure to reform and modernize itself throughout the 1960s, 1970s and early 1980s had led it to the edge of the abyss. Particularly implicated was the Brezhnev period (widely denounced at this time as an 'era of stagnation'), during which an ageing leadership had postponed needed changes, preferring a stable consensus to the turmoil of the past (Frankland, 1987; Sakwa, 1990, pp. 6–7). Gorbachev therefore came to aim at reform of the entire system, of what Zaslavsky called the 'neo-Stalinist state' (Zaslavsky, 1982).

Quite how this was to be done was another matter. No blueprint existed for successful reform of such a structure. At first, therefore, Gorbachev preferred to rely on tried and trusted methods like tightening up social discipline, especially discipline in the workplace. Most Western commentators are agreed that it was not until 1986 or even 1987 that the leadership became aware that the existing Soviet system would require more than mere tinkering if it were to survive and prosper. Even beyond that point, however, there was much official dithering about both the speed and direction of needed change.

Inevitably, Gorbachev's principal target was the neo-Stalinist command economy. Something will be said about his economic reforms, and the reasons for their failure, in chapter 4. Economic reform, however, was only one facet of what increasingly came to be seen as a many-sided attempt to de-Stalinize the Soviet system. Given the character of that system, any far-reaching economic reform was almost bound to involve the dismantling of the vast bureaucratic structure which lay at the heart of the command-administrative system. This in turn had serious implications for the careers, social status and privileges of the bureaucrats, politicians and generals who were part and parcel of that system.

Economic reform, therefore, inevitably became political and social reform, and this in turn engendered political struggle. Gorbachev's far reaching transformation of the political power structure, rescinding the Communist Party's legal monopoly on power, introducing neo-democratic initiatives like elections to the new assemblies (congresses and/or soviets) at national, republican and local levels and establishing a Soviet presidency (held by Gorbachev himself), plus the institution of many new civic and democratic liberties, were designed to head off his opponents in the party–state bureaucracy. Gorbachev was a sincere communist and had wanted to revitalize the system. The effect of his policies, however, was to undermine the position of the Communist Party.

The sweeping changes took place against a backcloth of mounting economic and political turmoil, reflecting the sheer difficulty if not impossibility of managing a smooth transition from a political and economic system which had been entrenched for more than sixty years. On the economic level, the many changes and dislocations began to send the country's economy into precipitate decline, with severe inflation, an alarming budget deficit, growing national debt, plummeting output, widespread shortages and rationing, the threat of mass unemployment, a booming black economy and much else. Economic difficulties were exacerbated by other problems in the social sphere, such as strikes, corruption, organized crime, ethnic strife and a general breakdown in law and order. And at the political level, the very cohesion of the country was called into question as republics, regions and even cities increasingly asserted their autonomy and their right to sort out their own problems locally. One very important facet of the situation was the way in which the entire record of the Communist Party's years in power, and thus the Soviet system itself, was increasingly criticized by a population made aware for the very first time of the many abuses and injustices perpetrated by the regime. As Richard Sakwa writes, until early 1990, 'Gorbachev's reforms were not designed to overthrow the old but to tap the potential of the existing system and to make it work better' (Sakwa, 1990, p. xi). After that time, the processes of change assumed a momentum of their own. Gorbachev's policy of *perestroyka*, implying managed change, had thus failed.

Ellman and Kontorovich have argued that, whatever may have been the underlying reasons for the failure of *perestroyka*, a considerable part of the blame for the collapse of the USSR can be laid at the door of the political leadership (Ellman and Kontorovich, 1992, pp. 1–39). In their view, Gorbachev made the supreme mistake of removing or weakening three crucial load-bearing 'bricks' in the structure he was trying to rebuild. These were: the central bureaucratic apparatus, the official Marxist-Leninist ideology and the party's active role in the economy. Without these 'bricks', the entire Soviet edifice was fatally damaged.

The assertion by the Union republics of their authority was a principal factor leading to the collapse of the Soviet Union. Republican leaders were able (or sometimes forced by their own populations) to build upon nationalistic feelings in a growing power struggle with Moscow. That they were in a position to do so was testimony to the failure of Soviet nationalities policy to bring about the merger between peoples which successive Soviet leaders had hoped for. The reasons for this failure and the rise in nationalist fervour towards the end of the 1980s have been examined by numerous writers (Smith, 1989; Kaiser, 1991, 1994; Zaslavsky, 1993). While the situation varied in every case, it is clear that national and ethnic strife was often fuelled by both long-term and short-term factors. Among the long-term factors, past injustices, the deterioration of the natural environment brought about by Soviet economic development policies and resentment of the social and cultural effects of Soviet nationalities and general development policies were frequently cited. Some scholars have argued that the very processes of modernization which characterized the Soviet period had themselves helped to foster national consciousness among the non-Russian peoples. The short-term factors included the continuing revelations about the many abuses which had been perpetrated by the Soviet system, and the economic and political mistakes of the Gorbachev leadership.

The rise of Russia was perhaps the most unexpected element in this situation. Following his election as chairman of the RSFSR Supreme Soviet in May 1990, Boris Yeltsin was able to assert Russia's distinct identity as against that of the USSR. In this he was supported by many liberals and democrats, frustrated by Gorbachev's numerous hesitations and compromises, and to some degree by various Russian nationalists, neo-communists and others, who resented what they saw as the undermining of Russian and Soviet power by Gorbachev and his associates. In June 1990, the RSFSR declared its sovereignty, meaning that it asserted the priority of its own laws over Soviet laws, and various territorial rights like rights to natural resources. Yeltsin was also able to furnish Russia with many of the attributes and symbols of statehood which it had previously lacked. This process culminated in June 1991 with Yeltsin's election as Russia's first president. In the meantime, both Gorbachev and Yeltsin strove to undermine each other's positions, Yeltsin speaking out for the rights of Union republics and various minorities, Gorbachev seeking to complicate the RSFSR's relations with its own republics and territories. Negotiations between the centre, the Union republics and other autonomous territories continued, with a view to finding a new basis for union. Eventually, Gorbachev's position was made untenable by the attempted communist coup in Moscow in August 1991. Gorbachev failed to place the blame for the coup attempt squarely on the communists, and most of the Union republics, fearing further coups, declared their independence. The independence of the three Baltic states of Estonia, Latvia and Lithuania was recognized by the USSR

Table 3.1 The 15 former Soviet republics: key territorial and economic indicators at the end of the Soviet period

	1	2	3	4	5	6	7	8
Russia	17,075.4	76.2	51.2	60.7	45.8	90.9	77.3	55.4
Ukraine	603.7	2.7	17.9	18.1	22.6	0.9	3.9	24.3
Belarus'	207.6	0.9	3.5	4.1	5.8	0.3	0.0	0.0
Uzbekistan	447.4	2.0	7.1	2.8	4.9	0.4	5.2	0.8
Kazakhstan	2,717.3	12.1	5.8	3.7	6.6	4.2	0.8	18.7
Georgia	69.7	0.3	1.9	1.5	1.5	0.0	0.0	0.2
Azerbaijan	86.6	0.4	2.5	1.7	2.0	2.2	1.4	0.0
Lithuania	65.2	0.3	1.3	1.4	2.2	0.0	0.0	0.0
Moldova	33.7	0.2	1.5	1.2	2.2	0.0	0.0	0.0
Latvia	64.5	0.3	0.9	1.2	1.4	0.0	0.0	0.0
Kyrgyzstan	198.5	0.9	1.5	0.7	1.2	0.0	0.0	0.5
Tajikistan	143.1	0.6	1.8	0.6	1.2	0.0	0.0	0.1
Armenia	29.8	0.1	1.2	1.0	0.7	0.0	0.0	0.0
Turkmenistan	488.1	2.2	1.3	0.5	1.1	1.0	11.3	0.0
Estonia	45.1	0.2	0.5	0.7	0.8	0.0	0.0	0.0

Key: 1, area (thousands of square kilometres); 2, percentage share of total USSR territory; 3, percentage share of total USSR population, 1991; 4, percentage share of USSR industrial output, 1985; 5, percentage share of USSR agricultural output, 1985; 6, percentage share of USSR oil and gas condensate production, 1989; 7, percentage share of USSR natural gas production, 1989; 8, percentage share of USSR coal production, 1989. *Source*: Shaw (1995, p. 152).

government in September. The final act came in December, when Boris Yeltsin and the leaders of other republics agreed to establish the Commonwealth of Independent States, consigning the USSR and the power of the Communist Party to oblivion. Henceforward, the Russian Federation, as it was now to be called in preference to the RSFSR, was left to make its own way in the world, shorn of its Soviet persona and separated from its 14 post-Soviet neighbours.

The Russian Federation Today: Ethnic and Political Geography

Despite being separated from the other former Union republics in 1991, the Russian Federation remains, as we have seen, the largest country in the world. It is easily the biggest of the post-Soviet states, with more than three-quarters of the territory and (in 1991) over half of the population of the former USSR (table 3.1). It also overshadowed the other republics in terms of industrial output and its share of natural resources. Only in agricultural output was its lead rather more modest.

Ethnically, Russians dominate the Federation, with over 80 per cent of the total population of 147 million in 1989. This means that the titular population in Russia has a higher percentage share of the total population than in 12 of the 15 post-Soviet states. Even so, the total number of ethnic non-Russians living in Russia in 1989 was far from insignificant, at more than 27 millions. Russia is ethnically very diverse. According to the handbook *National Composition of the USSR Population*, based on the 1989 census, there are 67 numerically significant nationalities if the minority peoples of the north are counted as one group, or 92 if the northern peoples are counted separately (*Natsional'nyy sostav*, 1991, pp. 28–33). The Slavs (Russians, Ukrainians, Belarusians) clearly predominate, with over 85 per cent of the total (table 3.2). Other groups include the various Turkic and related peoples living in an area stretching from the Volga region and towards the Urals (Tatars, Bashkirs and the Hunnic Chuvash) and scattered across Siberia (Yakuts or Sakha, Tuvinians and others), the Mongol peoples (notably the Buryats in East Siberia and the Kalmyks living on the north-western shore of the Caspian Sea), the Caucasians (living in the North Caucasus) and the Finno-Ugrians (who include the Mordvinians, Udmurt and Mari of the Volga region, the Komi of northeast European Russia and the Karelians living near the Finnish frontier). Jews and Germans are also numerically important, the latter mainly descended from eighteenth- and nineteenth-century migrants who were invited to Russia to settle the steppe. The northern minority peoples, though numerically insignificant (about 180,000 in 1989), are a bewildering variety of Finno-Ugrian, Altaic and Palaeoasiatic groups (the latter including the Chukchi and Koryaki of the north-east), with only a handful of Inuits (*Eskimosy*) and Aleuts living on the easternmost extremity of Russian territory. The complex ethnic geography is depicted in figure 3.1. With an influx of Russians moving into the Russian Federation from the other post-Soviet states since 1989, and an outflow of other nationalities (Ukrainians, Belarusians, Germans, Jews and others) the detailed picture will now differ somewhat from that shown in table 3.2.

It is worth noting that there are some discrepancies between what a person may consider his or her nationality or ethnic identity to be, and the official designation. From the early 1930s (later in the case of peasants), Soviet citizens were obliged to carry internal passports in which their ethnic identity was recorded. This designation was inherited, although children of mixed marriages had a choice when they came to be registered for a passport. By one means or another many people managed to be registered as Russians because this was considered more

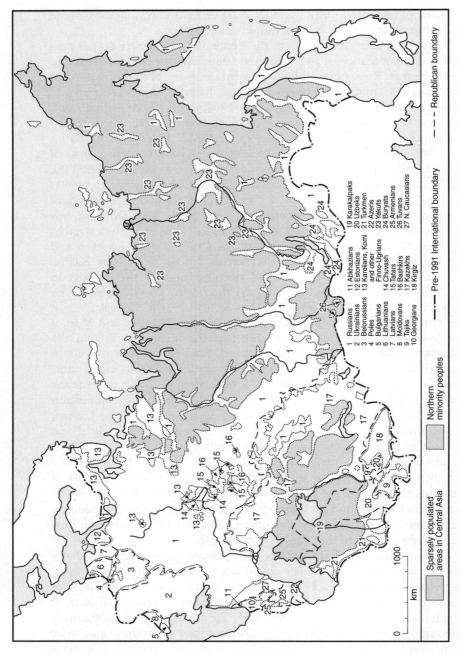

Figure 3.1 Ethnic geography of Russia and the successor states.

1 Russians
2 Ukrainians
3 Belorussians
4 Poles
5 Bulgarians
6 Lithuanians
7 Latvians
8 Moldovans
9 Tajiks
10 Georgians
11 Abkhazians
12 Estonians
13 Karelians, Komi and other Finno-Ugrians
14 Chuvash
15 Tatars
16 Bashkirs
17 Kazakhs
18 Kirgiz
19 Karakalpaks
20 Uzbeks
21 Turkmen
22 Azeris
23 Yakuts
24 Buryats
25 Armenians
26 Tuvans
27 N. Caucasians

Sparsely populated areas in Central Asia

Northern minority peoples

Pre-1991 International boundary

--- Republican boundary

0   1000   km

Table 3.2   Major nationalities in the Russian Federation, 1989

Nationality	Number (thousands)	Percentage of total RF popn
Russians	119,866	81.5
Tatars	5,522	3.8
Ukrainians	4,363	3.0
Chuvash	1,774	1.2
Bashkirs	1,345	0.9
Belarusians	1,206	0.8
Mordvinians	1,073	0.7
Chechens	899	0.6
Germans	842	0.6
Udmurts	715	0.5
Mari	644	0.4
Kazakhs	636	0.4
Avars	544	0.4
Jews	537	0.4
Armenians	532	0.4
Buryats	417	0.3
Osetians	402	0.3
Kabardinians	386	0.3
Yakuts (Sakha)	380	0.3
Dargins	353	0.2
Komi	336	0.2
Azeris	336	0.2
Kumyks	277	0.2
Lezgins	257	0.2
Ingush	215	0.1
Tyvans	206	0.1
Peoples of the North	182	0.1

Source: 1989 Census of Nationality.

prestigious. Nowadays, however, this may no longer apply, and in future censuses people may change their ethnic identity back to their ancestral one. Other anomalies also occur. Thus numerous small nationality groupings were not officially recognized, or ceased to be recognized at a particular point in time. Where ethnic groups had or have no official homeland, their status was always an uncertain one. Many ethnic groups, of course, lack an official homeland and, since Stalin's abolition of the national soviets in the late 1930s, have had no status at all in the administrative hierarchy.

The federalization of the RSFSR during the Soviet period has been discussed above. In 1989 the Russian Federation contained 16 autonomous republics (ASSRs), five autonomous oblasts and ten autonomous okrugs. The federal structure inherited from the Soviet period has been retained, but there have been a number of changes, including name changes (see figures 1.1 and 3.2). All sixteen of the autonomous republics now have the status of 'republics of the Russian Federation', implying a claim to sovereignty (even though this is not recognized by the 1993 constitution) and more autonomy than they enjoyed as autonomous republics. Four of the five autonomous oblasts have also been raised to republican status, the Jewish Autonomous Oblast in the Far East being the sole exception. Moreover, the former Chechen-Ingush ASSR in the North Caucasus has split into two republics: Chechnya and Ingushetia. There are therefore now 21 republics within the Russian Federation, one autonomous oblast and ten autonomous okrugs. In addition to the republics and other autonomous entities, the Federation contains 57 'non-autonomous' administrative units, namely the ordinary oblasts (regions) and krays (territories), plus the cities of Moscow and St Petersburg. Altogether, then, the Federation consists of 89 regional administrative entities. Oblasts and krays are commonly subdivided into rayons (districts). The status of these various entities, and their evolving relationship with the central government, must now be considered.

The Russian Federation and Problems of Federalism

The circumstances in which the Russian Federation asserted its independent identity and helped to dispose of the USSR hardly contributed to political stability. As we have seen, Gorbachev and Yeltsin had engaged in a struggle for political power by trying to obtain the backing of Union republics, the lesser ethno-administrative units and ethnic minorities. The effect of *perestroyka* and of its accompanying political struggles had been to call into question not only the unity between the fifteen Union republics but also that with the lesser ethnic units. Thus, Russia's assertion of its sovereignty in June 1990 was quickly followed by similar declarations on the part of its (autonomous) republics as well as by some of the autonomous oblasts and okrugs. A few of the latter also made bids for full republican status. The prime target of all these proclamations was the Soviet government in Moscow – in other words, the old Soviet system. With the fall of the Soviet system in 1991, however, the target for Russia's republics and autonomous regions became the Yeltsin government. The Russian Federation seemed in danger of breaking up, just as the USSR had done.

The fact that the Union republics had declared their independence was not entirely surprising. As we have seen, several of them had known periods of independence during the civil war or later. The republics

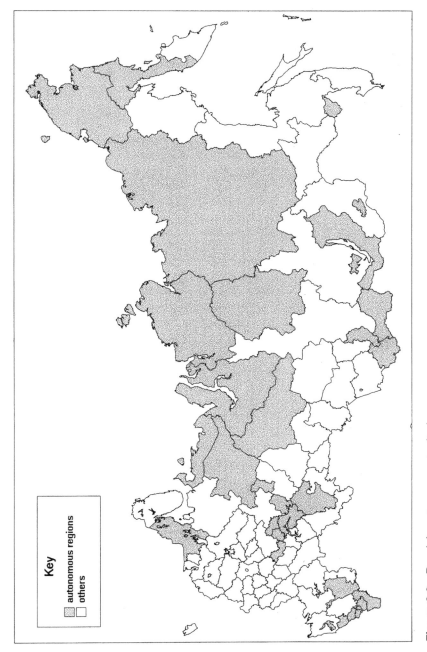

Figure 3.2 Russia's autonomous territories.

Key

autonomous regions

others

which had founded the USSR by treaty in 1922 had an historical claim to sovereignty based upon that fact. Moreover, the myth that the USSR was a voluntary union between Union republics seemed to give all of them a legitimate right to an independent existence. The freedom to secede from the USSR, should any republic wish to do so, had been enshrined in the 1936 and 1977 Soviet constitutions. During the Soviet period, as noted above, Union republics enjoyed many of the trappings of statehood. This even included the conduct of foreign relations (for example, Ukraine and Belarus', as well as the USSR, were members of the United Nations, although always voting with Moscow). We have also seen that Soviet nationalities policy did not succeed in eradicating national differences and may even have reinforced them in practice (Smith, 1985; Zaslavsky, 1993). For the Union republics to declare their independence once the Soviet system had manifestly failed (and especially in the light of the threat to their newly won rights posed by the August 1991 coup attempt) was perfectly understandable in the circumstances.

The position of the (autonomous) republics and smaller ethnic units within the Russian Federation was less straightforward. These had barely known independence in the modern period. Furthermore, most had been created by central fiat during the 1920s and 1930s, and they thus lacked the historical legitimacy of the Union republics. Other complications included the fact that many of them had a majority Russian population, that they were sometimes geographically positioned in the midst of Russian territory and that cultural and historical links with Russia were sometimes close (see chapter 9). Even so, according to some scholars, the mere fact that ethnic groups had been granted their own territories by the Soviets was an important factor behind the growth of national identity (Smith, 1991). Once it became possible, under Gorbachev, for Union republican and other regional leaders to accuse the USSR of having practised imperialism against them, it became a simple matter for minorities within the Russian Federation to level similar charges against Moscow. Just as the USSR was increasingly portrayed before 1991 as an empire in decay, so many of the Russian Federation's non-Russian citizens were tempted to view the Federation in the same way.

After 1991, therefore, the stage was set for a struggle between Moscow and the republics and other autonomous entities within the Russian Federation. The situation was made particularly difficult by the power vacuum at every level of government which had been left by the collapse of communism. Thus, although popular elections in 1990 and 1991 had provided some democratic mandate for the powers now being exercised by the Russian president and parliament (then known as the Congress of People's Deputies), there was no established constitution or credible legal system to determine how those powers were to be used. The political process was still regulated by the old RSFSR constitution of the Brezhnev era, which was clearly out of date in a post-communist age. Lack of respect for this legal framework, coupled with a general absence of the workings of a democratic polity, quickly threatened to generate anarchy. The most striking example of this problem was the struggle for power between President Yeltsin and the Congress of People's Deputies. The struggle between executive and legislative authorities was repeated in republican, regional and local authorities throughout Russia. Added to this was the struggle between the centre and the regions. Thus the republic of Checheno-Ingushetia declared its independence from Russia towards the end of 1991, and the republic of Tatarstan's referendum on sovereignty in the following March virtually amounted to the same thing. Other republics and autonomous units made varying claims to enhance their rights and their freedom. Arguments raged over such issues as property rights, ownership of natural resources, taxation, linguistic and cultural freedoms. Meanwhile, non-ethnic units of the Federation, jealous of some of the rights being claimed by autonomous units and also keen to bolster their own freedom and development prospects, began claiming enhanced status and banding themselves together into various kinds of regional grouping. Much of this was in reaction to both the economic and political chaos which had only been exacerbated by the final collapse of the Soviet Union. Fragmentation of the Russian Federation seemed likely to follow hard on the heels of that of the USSR.

Unfortunately, the experience of federalism during the Soviet period provided little guidance for the reconstruction of the Russian Federation on a new basis. Scholars have classified the world's federations into a number of types (Wheare, 1963; Smith, 1995) but most are agreed that the USSR was not really a federation at all. It is usually argued that the concept of a federation implies a clear and generally agreed division of powers between two levels of government, as in the United States (Paddison, 1983, p. ix). But in this sense the USSR had only the appearance of a federation. In reality, political power was strongly centralized by the Communist Party and the central authorities exercised the right to intervene in republican and regional affairs to any extent they deemed necessary. In practice, the various USSR and republican constitutions could only be enforced by the Communist Party, which also felt free to violate them at will. The right of secession accorded to Union republics was a myth.

Soviet history, however, provided one precedent which those hoping to preserve the Russian Federation regarded as useful. This was the

Union Treaty of 1922, binding the four original Union republics into the USSR (see chapter 2). Whatever may have been the political reality behind that event, at least it purported to show that the union was originally a voluntary one. It was a precedent which Gorbachev had hoped to follow in 1991 in negotiations with the Union republics for a Federation Treaty. This would have re-established the USSR on a new, agreed basis. In the event, the endeavour came to grief as a result of the failed August 1991 coup. After the fall of the USSR, President Yeltsin similarly found himself negotiating for a treaty with Russia's republics. Ironically, the RSFSR had never been a treaty-based federation during the Soviet period, thus differing in its constitutional history from the Soviet Union itself.

In March 1992, Yeltsin managed to induce 18 of the then 20 Union republics to sign a new Federation Treaty with the federal government. This was at a time when Yeltsin was hoping to outmanoeuvre his political opponents in the Russian parliament. The treaty thus demarcated between the powers of the federal government and those of the republics, granting new powers to the latter. At the same time, many of the most contentious issues, such as the ownership of natural resources and taxation principles, were left unspecified (Shaw, 1992). Treaties were also signed with the other autonomous entities within the Russian Federation and with the non-autonomous units. On paper, at least, the Federation had thus been reconstructed on a treaty basis.

These agreements failed to provide political stability, being fatally undermined by the continuing struggle between president and parliament. Ominously, for example, neither Tatarstan nor Checheno-Ingushetia was party to the Federation Treaty, while several republics adopted constitutional provisions which violated Russian Federation laws. Tatarstan regarded itself as only 'associated' with the Russian Federation, implying that the latter could only exercise powers in Tatarstan voluntarily ceded by the Tatar government. The Republic of Tyva in Siberia claimed the right of secession, contrary to Russian law. Other republics, autonomous areas and ordinary regions competed for power both with the centre and with one another in a seemingly unstoppable process.

Finally, in September 1993, Yeltsin moved to put a stop to the political turmoil by dissolving parliament and ordering elections for a new Federal Assembly in December. The presidential decree was resisted both by the legislature and on the streets, as a result of which Yeltsin used military force to storm the parliament building and arrest his political opponents. Many liberals and democrats were dismayed by these events, regarding them as a reversion to older traditions of Russian authoritarianism and a regrettable precedent in the context of infant Russian democracy. The December elections brought into being a new and weaker Federal Assembly, which henceforth was to consist of a lower house, the State Duma (so named in deference to the pre-1917 parliament), and an upper house, the Federation Council. The election results were a disappointment to President Yeltsin and his supporters, since they ensured the entry to the new parliament of many conservative nationalists and communists, who were opposed to the president's economic and political policies. However, the results of a national referendum on a new constitution held at the same time allowed the president to claim sufficient support for his determination to reconstruct both the central government and the balance of power between centre and periphery.

The Russian Federation which has emerged as a result of the December 1993 constitution is a more centralized polity than that which immediately preceded it. Wide powers are now exercised by the president, and the ability of parliament to interfere with his actions or those of his government is much reduced. All 89 administrative regions are now regarded equally as 'subjects' of the Federation, and no rights of secession or declarations of sovereignty are recognized by the constitution. The republics are, however, still referred to as 'states' and have other symbolic rights like constitutions of their own. The 1992 Federation Treaty has not been adopted as an annex to the constitution, contrary to earlier promises. Its exact legal status thus remains uncertain, reflecting the reduced powers of the republics under the new dispensation.

That President Yeltsin was able to strengthen his powers in this way is a reflection of increased support from, and reliance upon, powerful vested interests, especially the military. The growing authoritarianism of the Russian government after 1993 was noted by many, together with its apparent determination to hold the Federation together. Thus, during the course of 1994, new treaties were signed with several of the republics and economic pressure was brought to bear upon Tatarstan, for example, to bring it into line. New legislation now controls such matters as the division of revenues from taxation between federal and regional authorities, though republics, because of differences in their resource endowment and political influence, are still treated unequally. Russia has thus been described as an 'asymmetrical federation' (Lynn, 1996). Yeltsin also placed increased reliance upon his appointed regional governors to control the periphery, though these were subject to popular election from 1996. The most spectacular instance to date of the determination to preserve the Federation has been the bloody Chechen war, launched in December 1994. The assault upon the breakaway Caucasian republic appalled many by its ferocity and led numerous commentators to argue that Russia was reverting once again to older and nastier traditions of brutality and repression.

If the term federation implies a definite and enforceable division of powers between two levels of government, then the Russian Federation, with its presidential powers and flexible constitutional traditions (including the lack of an effective constitutional court), falls short of that definition. It may be questioned whether the final political balance between centre and periphery has yet been reached. On the one hand, the Russian state seems determined to enforce its territoriality and there are many voices calling for reconsideration of the present administrative structure: for example, because of the threats it poses to the unity of the state, or because it is a hindrance to the success of economic reform. The major candidates opposing Yeltsin in the June 1996 presidential election – the communist Zyuganov and the nationalist Zhirinovsky – would certainly have taken a hard line with the republics if they had won. On the other hand, republics are still in the business of nation building within their own territories, providing themselves with presidential systems and other symbols of nationhood and attempting to secure enhanced economic, political and cultural control over their own affairs. Many republican constitutions continue to contradict the federal one in various ways (Magnusson, 1996). The 1993 federal constitution itself still leaves much to be decided about the powers of republican and regional authorities, and about the relationships between them. The continued propensity of the current federal government to negotiate treaties with republics and power-sharing agreements with regions suggests that the Federation is still evolving, perhaps in a rather *ad hoc* way.

It also needs to be emphasized that struggles for power between the centre and the regions have now become a characteristic of states across the world, and that Russia is thus by no means unique in this. One feature of Russian regional politics in recent years has been a tendency for regions to cultivate their own international relations, to some extent by-passing the centre. This too is an international phenomenon, as regions seek inward investment from the global economy. It is clear that in the future Russian regions will have to take on much greater responsibility for their own economic development. States across the world are thus witnessing the development of sub-national actors with their own international agendas – what some political scientists have described as a trend towards 'perforated sovereignty' and 'marbled diplomacy' (Hughes, 1996).

A number of authors have pointed to factors which may continue to promote instability within the Russian Federation (Walker, 1992; Shaw, 1993). These include: a lack of legitimacy accorded to the Federation by

its ethnic minorities (particularly under conditions of intensified ethnic
nationalism); lack of public experience of and respect for constitutional-
ism and the sanctity of law (an old Russian trait which even finds
expression in some of the actions of the present government); the only
partial character of democratization (with some of the centrifugal forces
being fomented by local ex-communist elites, still in power, determined
to maintain their authority and their privileges in the new circumstances); and troubles arising out of the conflicting claims of ethnicity
and individual liberty. There is also the crucial question of whether
the post-1991 Russian Federation, now separated from the peripheral
territories which were occupied by the Russian Empire and the Soviet
Union, can sufficiently satisfy the national aspirations of ethnic Russians.
This issue must now be addressed.

The Russian Federation and Russian National Identity

Many Russians, even some who are anti-communist and not particularly
nationalistic, have bemoaned the collapse of the USSR as a unified state.
Two factors seem to lie behind this. One is the sense of loss involved in
being cut off from territories having a long association with the Russian
state (Dunlop, 1993). The other is the realization by Russians that they
are no longer the citizens of a superpower. The latter is a feeling of
bewilderment which has been felt by others, notably by numerous citizens of Britain and France with the disappearance of their empires after
the Second World War. It could be argued that the policies of both
countries have since been strongly influenced by the need to come to
terms with this loss. The sense of loss experienced by Russians may be
even greater. Geography determined that there was a definite territorial
divide between the west European states and their overseas colonies, and
this divide was reinforced by their administrative arrangements (Zubov,
1994). No such geographical and administrative divide existed in the
Russian case. The old Russian Empire was by and large administered in
a unified way, and little attempt was made to distinguish between
the ethnically more and less 'Russian' parts of the state. After 1917 the
Soviets did subdivide the territory, as we have seen, but the borders of
the RSFSR had little significance for any Russian sense of national
identity.

Today, then, the majority of ethnic Russians find themselves living in
a state, the Russian Federation, which had little or no meaning for most
of them before 1991. Not surprisingly, this has led to debates about the
exact nature of this state and its historical legitimacy. For example,
questions have been raised over whether this multinational state is meant

to be representative of all the peoples now living within it, or whether it is primarily a nation state representing the ethnic Russian majority. At one level, it is clear that the former is supposed to be the case, since, alone among the post-Soviet states, Russia is a federation. At another, however, there seems to be some doubt about the matter, since the 1993 constitution names the state both 'Russia' and the 'Russian Federation'. Questions have also been raised about the way the Russian Federation came into being originally. Some anti-communists, for example, have argued that the RSFSR was created in an arbitrary and purely opportunistic way by the Bolsheviks back in the early Soviet period, and that such decisions need not be taken as binding today. As one commentator has put it, 'the RSFSR has neither historical nor ethnic legitimacy' (quoted in Walker, 1992, p. 15). The mirror image of this argument is the one put forward by many communists, namely that the decision to break up the USSR in 1991 was illegal and should therefore be reversed. The latter opinion was endorsed by a decision of the communist and nationalist-dominated Russian parliament in March 1996.

A closely related problem is that of borders. As noted earlier, the RSFSR's borders were originally defined early in the Soviet period in such a way as to leave many ethnic Russians living in other republics. As long as the USSR endured, this hardly mattered and many ethnic Russians moved to the other republics after 1917 to take up jobs there. As a result, by 1989 some 25 million Russians (or over 17 per cent of all the Russians in the USSR) were living outside the Russian Federation. Nowadays such erstwhile migrants, most of whom have not yet returned to Russia, find themselves minorities in independent states. Sometimes they find themselves discriminated against in favour of the titular nationalities (see chapter 10). Needless to say, this situation fuels the emotions of those Russian nationalists who would like to see Russia's borders expanded to incorporate such peoples.

The question of borders affects Russia's relations with its post-Soviet neighbours, and this too is an issue which worries some Russian nationalists. As noted in chapter 1, Ukraine and Belarus' are Slavic countries having close historical and cultural ties with Russia. The celebrated novelist Alexander Solzhenitsyn has lamented their separation from Russia as essentially artificial, and has called for reunification within a revived Slavic commonwealth (see Solzhenitsyn, 1991).

The dispersal of Russian settlement across the territory of the former USSR means that almost any Russian Federation border would be open to dispute and makes the attempt to define a truly Russian homeland a hopeless one. In political geography, the 'homeland' is that territory with which a group of people, constituting a 'nation', feels a particularly close historic link and attachment (Smith, 1991; Taylor, 1993, pp. 196–7). In the Russian case, the 'homeland' is sometimes taken to be that territory of central and northern European Russia which was settled by the Russians prior to the period of imperial expansion inaugurated by the conquest of the Tatar state of Kazan' on the Volga in 1552. This was the sense in which it was used in chapter 1. This was also the area which, in the 1970s, was designated a 'non-Black Earth' development zone and which was felt by many nationalists to be the core of ethnic Russia (Vanderheide, 1980). The landscapes of this region may thus have something of the same symbolic significance for Russian national identity which the rural landscapes of southern England or of New England have in their respective countries (Lowenthal and Prince, 1965; Meinig, 1979). But the problem is that this is only the 'original' Russian homeland. It provides no guidance whatever about the rightful extent of the Russian state today.

Uncertainties about the status of the Russian Federation no doubt relate to deep-rooted insecurities of Russian national identity, insecurities which have been exposed by the sudden disappearance of the USSR. The history of Russia, for example, provides scant guidance about what Russia is meant to be today. In the case of some other countries, significant historical events, like the American Declaration of Independence and the French Revolution, or long-standing institutions like the British monarchy, have seemed to provide a sense of continuity and meaning to the nation. But Russian history has been so turbulent and disjointed that it provides few signposts for present-day Russians to follow. Of course, the importance of language, culture and even religion must not be underestimated, and many Russian nationalists have hoped that they might provide a solid bedrock for a new national identity to be built upon. But both culture and religion suffered at the hands of the communists and, though making use of Russian nationalist feeling for their own ends, the communists did much to erode a sense of Russian distinctiveness in practice.

The task which the leaders of the new Russia have set themselves, therefore, is to build a sense of identity which will somehow hold the vast and multi-ethnic Russian Federation together, preferably without involving it in conflict with the outside world. Russian ethnicity will obviously be important, but a too narrow and exclusive reliance on that will run the risk of alienating the many ethnic minorities. Equally, policies which favour some regions at the expense of others will run similar risks, and these may be exacerbated by claims to regional autonomy (examples might include Siberia, and the cossacks of the Russian south). All this is ultimately linked to the historic conflict over whether Russia should seek to modernize following a Western or other foreign model, or whether it should seek its own distinctive way in the world. In other words, Russian national identity is intimately linked with the future development of Russian society and with issues to be raised in the chapters that follow.

Four

The Command Economy and the Transition to Capitalism

Since the collapse of the USSR towards the end of 1991, Russia has been struggling to transform itself into a market economy. That struggle has been far from easy, and whether the country can achieve the economic and political stability its population is hoping for remains in some doubt. The problems and repercussions of economic transition are discussed below. But before they can be considered, more must be said about the command economy which dominated the Soviet Union for so long. Unless the nature and purpose of the command economy are understood, the difficulties which have attended the attempt to marketize can scarcely be appreciated.

The Command Economy in the Post-Stalin Period

It is important to understand that the command economy was designed to fulfil different purposes from the market economy, however much the functioning of the latter may be influenced by political, military and other considerations. Stalin wished to build up the USSR's heavy industrial might with the purpose of enhancing its military power. Industrialization created a vast military machine which secured a great victory in 1945 and which promised to defend the USSR and its allies in the competition with capitalism thereafter. The USSR's international ambitions were undoubtedly excited by the successes it enjoyed in the latter part of the Second World War and its immediate aftermath, and the command economy became the foundation for the 'military–industrial complex', or what has sometimes been referred to as a 'permanent arms economy' (see Cliff, 1964, 1974). Military might required economic growth, and the country's major industries justified themselves in terms of the needs of the military. It was a self-reinforcing mechanism

Table 4.1   Annual average growth rates of real
GNP, 1953–65, USSR and major capitalist states

	per employed person (%)	per man-hour
USSR	3.8	5.2
France	4.6	4.6
West Germany	4.6	5.4
Italy	4.3	4.5
USA	2.3	2.2
UK	2.4	2.7
Japan	7.7	7.8

Source: Munting (1982, p. 138).

which moulded the character of the economy down to the end of the
Soviet era.

Of course, the command economy also had other purposes. One of
them, which became more prominent after Stalin's death, was to prove
the superiority of socialism over capitalism by overtaking the West
economically. Soviet leaders long expressed the view that it was only a
matter of time before this would happen, and then their people might
enjoy higher living standards and better welfare provision than was
possible under capitalism. In the meantime, the command economy had
the important task of securing the Communist Party's political power
base by raising living standards generally and, above all, by rewarding the
party's most trusted supporters and key workers. The 'new class' (Diljas,
1966) of party functionaries, administrators and technocrats expected
both material and social privileges in return for their loyalty.

The command economy enjoyed reasonable success during the Stalin
years (when there was considerable reliance on forced labour) and
through most of the 1950s. After that time, however, the annual growth
rate slowed to a level which was fairly close to that of several
major capitalist states (Munting, 1982, p. 138) (table 4.1; see also tables
2.2, 2.3). The worrying thing for the Soviet leaders was that this decel-
eration was not accompanied by any great diversion of income to consumption, as had occurred in the West. This led many to question,
perhaps for the very first time, whether the goal of catching up with
and overtaking the West was going to prove more problematic than
previously thought.

The response to such difficulties took a number of different forms.
Because of the unbalanced character of the economy inherited from
Stalin, a lack of balance reflected in the over-emphasis on heavy industry

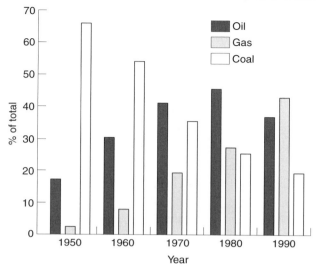

Figure 4.1   The Soviet fuel balance, 1950–90.

and thus on industrial regions which had first developed in the late nineteenth century, the post-Stalin leadership felt obliged to give rather more attention to relatively neglected sectors like agriculture, transportation, the consumer sector and social welfare. But, because their priorities had not really changed from those of Stalin, and because of the vested interests which had built up during his period, they decided to do this without any absolute downgrading of the importance of heavy industry and the military. Ultimately, this meant that the command economy retained much of its Stalinist character. The problems this posed for such sectors as agriculture and the consumer sphere will be explored in later chapters of this book.

Another response was to try to make Soviet industry more efficient through a concerted modernization programme. For example, it became apparent in the late 1950s that the USSR was lagging behind the West in the substitution of plastics for more traditional materials and in the production of certain types of chemicals. The economy was in fact still oriented to traditional industrial fuels like coal and to metals like iron and steel. Thus there followed a restructuring of the country's fuel balance in favour of oil and natural gas, with their many advantages in terms of costs of exploitation and transport (figure 4.1). Geographically,

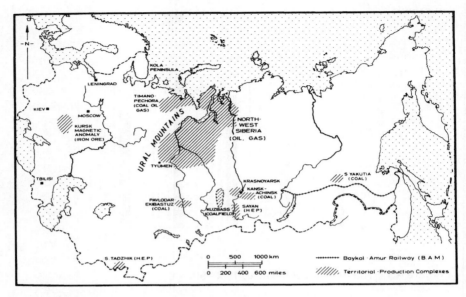

Figure 4.2   Major regional developments in the post-Stalin era.

this led to growing investment in the conveniently located energy sources in the European territory, but also increasingly in peripheral regions. Thus by 1980, West Siberia was supplying half of the USSR's oil and almost a third of its natural gas. The westward movement of huge quantities of fuels from Siberia was made possible only by the growing pipeline network. This network in turn allowed the dispersal of new industries away from traditional centres to secure local advantages like labour surpluses or the use of transportation arteries. The dispersal was also fostered by the construction of integrated electric power grids. Thus along the Volga and towards the Urals there arose new chemical and petrochemical industries to supply the nation's growing need for fertilizers, plastics and other materials. Much the same began to happen in the southern part of West Siberia, prompting one Western geographer to talk of a 'new Soviet (industrial) heartland' (Hooson, 1964) (figure 4.2).

Similarly, new investment was now being poured into modernizing the metallurgical industries to raise quality and to produce new metals needed by the electronic, aerospace, nuclear and other industries: hence the development from the 1960s of the ores and associated industries of the Kursk Magnetic Anomaly in central European Russia. Investment also went into new branches of the machine building and engineering industries, like instrument making, electronics, computers, telecommunications and other high-technology industries. These developed around such traditional engineering centres as Moscow or Leningrad (St Petersburg), as well as in more dispersed locations, like the capitals of the Baltic republics, with a tradition in specialized engineering. Motor car manufacture also received a boost from the 1960s, and several major new plants appeared in the rapidly developing Volga region.

The continued industrial development of the European territory was fostered not only by its inherent advantages in climate and population but also by the growing links with the COMECON countries of eastern Europe and, after about 1970, with the West. But industrialization of the European regions also brought numerous problems, such as a growing energy deficit, water shortages and severe pollution. Development of the resources and associated industries of the eastern regions beyond the Urals particularly focused on the 'territorial production complexes' (Linge et al., 1978: see figure 4.2). But here industrial development encountered problems like high environmental costs and labour shortages. Thus even by 1988, the eastern territories (excluding the Urals) still accounted for only 22.9 per cent of Soviet industrial production, compared with 8.3 per cent in 1940 (Dmitrieva, 1996, p. 39; cf. Adamesku and Kistanov, 1990, p. 112).

Other responses to the problem of slowing economic growth rates included various attempts at economic reform (trying to make planning more efficient and introducing some market-type elements into the system), pursuing policies of détente with the West (the rising costs of the military machine proved a real burden on the economy) and importing more Western technology. None of these responses proved especially effective, for reasons to be discussed below.

Problems of the Command Economy

Tables 2.2 and 2.3 suggest that Soviet economic performance began to lag seriously behind expectations, especially after 1970. In other words, despite the many changes which had been implemented in the post-Stalin years, the economy's ability to fulfil the goals outlined towards the beginning of this chapter were called into question. By the 1980s, many commentators were beginning to feel that only radical economic reform could redeem the situation.

Why did the command economy fail in this fashion? By way of an answer, we can divide the problems encountered by the Soviet economy into sectoral and geographical ones, although the distinction is a rather artificial one.

As regards *sectoral* problems, these have been extensively analysed by various writers (Nove, 1987; Hewett, 1988; Dyker, 1992; Gregory and Stuart, 1994) and only a few salient points need be made here. It has already been suggested that the economy became increasingly complex in the post-Stalin period. Partly this was because of the need to achieve a better balance, since neglecting agriculture, transportation and other areas was a policy which could not be pursued for ever, without encountering considerable economic and social costs. In agriculture, for example, poor performance (for reasons to be discussed in chapter 7) engendered the need to rely increasingly on food imports, a real burden on the economy. The economy also became more complex as a result of the necessity to raise living standards to retain the political support, or at least the acquiescence, of the masses and to deflect unfavourable comparisons which might be made with the West. The problem was that a more complex economy, and a more demanding population, proved ever more difficult for the planners to cope with. The growing bureaucracy was unwieldy and stifled incentives and innovation. Despite such intiatives as trying to introduce economic reforms, increase incentives and take advantage of modern technologies like computers, the gap with the West seemed only to grow larger. The Soviet command economy simply failed to emulate the flexibility and inventiveness associated with the market.

Geographical problems can be subdivided into several types.

Numerous problems flowed from the *spatial behaviour of the industrial ministries* which were responsible for administering the various sectors of the command economy during most of its existence. Since the central planners in *Gosplan* were incapable of making many detailed decisions for each individual sector, the ministries were left with considerable discretion: for example, over locational decisions. In making their investments, ministries tended to play safe and choose locations which were most convenient for themselves or where they knew that their enterprises would be able to secure their needed raw materials, supplies and essential services. In practice, this often meant big cities. From the economic point of view, the locational decisions made by the ministries were not necessarily the cheapest ones or the best from the perspective of regional planning. In any case, such decisions often encouraged excessive immigration into big cities, leading to pressure on housing, services and the environment. Because they favoured key locations and regions, ministerial decisions often exacerbated regional inequalities. A further problem lay in the tendency for ministries to avoid having to rely on one another more than was necessary. This classic bureaucratic behaviour

Table 4.2 Proportion of Soviet oil and natural gas production contributed by Russia and by Siberia and the Far East, 1970/5–89 (percentages)

	1970	1975	1980	1985	1989
Oil production					
Russia		83.3	90.6	91.1	91.0
Siberia/Far East		30.7	52.3	62.3	67.3
Natural gas production					
Russia	42.1	39.8	58.3	71.8	77.4
Siberia/Far East	5.3	13.7	37.2	59.5	68.1

Source: Sagers, 1990, pp. 281, 290.

(often referred to by Soviet commentators as 'narrow departmentalism') derived from the unreliability of inter-ministerial supplies, a problem which always plagued the command economy. It therefore made sense for ministries to be as self-sufficient as possible, but this led to much waste and duplication of effort, while lack of coordination between ministries, and between ministries on the one hand and regional or urban authorities on the other, constantly threatened to undermine urban and regional planning (Shaw, 1985, 1986).

A further characteristic set of problems of the command economy related to *resource use*. The USSR, for example, was a major energy producer, supplying not only its own needs but also those of its allies in COMECON. After 1970, the country came to rely more and more on energy exports to the non-communist world to earn the hard currency needed to pay for the technologies, food and other necessities it was unable to produce itself. The energy industries were therefore under considerable pressure and over time, as the energy reserves of the European territory were run down, greater reliance had to be placed on eastern resources (figure 4.2). This meant high costs of development in regions with difficult environmental conditions, such as the swamplands of West Siberia, which were producing over two-thirds of Soviet oil by 1989, and the harsh and remote coastal tundra in the north, where over two-thirds of natural gas were being produced by the same year (table 4.2). Even in coal, a relatively neglected sector, it became necessary to place greater emphasis on the eastern fields, with those beyond the Urals accounting for over 60 per cent of production by 1989. In addition to the high costs of development, the Soviet economy had to bear the costs of investment in oil and gas pipelines and of transporting the various forms of energy to where they were needed, often huge distances away. Alternative policies, such as developing the available energy reserves in the European regions, proved even more expensive and, in the case of nuclear energy, carried enormous risks, as the catastrophic accident at Chornobyl' in 1986 illustrated only too well. Moving energy-demanding industries to the east was also expensive and resisted by the conservative ministries. Finally, the command economy had a very poor record when it came to energy conservation, since ministries and their enterprises had little incentive to save on energy in the absence of adequate administrative or financial pressures.

The USSR thus remained an energy-intensive economy, but the geography of energy production meant that costs were constantly rising. By the 1980s, for example, the energy sector was absorbing almost 20 per cent of investment capital (Gustafson, 1989, p. 47). The point was that money invested in energy was not then available for other pressing needs, such as industrial modernization, agriculture or the consumer sphere. The energy sector thus came to act as a serious brake on Soviet economic performance – 'one of the leading proximate causes of the downturn and stagnation of Soviet economic growth' (Gustafson, 1989, p. 5).

Rather similar problems afflicted other kinds of natural resource use. Excessive demand led to shortages (as in water – see chapter 6), and to the need to invest heavily in inferior resources located in the European territory (as in ferrous metallurgy) or to turn to more abundant reserves in the periphery. An exploitative attitude, particularly characteristic of low priority areas like forestry, often meant sacrificing long-term prospects for short-term gains (Barr and Braden, 1988).

Just as the command economy was lavish in its use of natural resources, so it was with *labour*. Unlike in many market economies, where market pressures have proved potent in forcing firms to minimize labour costs, the Soviet economy, with its artificial price structure and inbuilt conservatism, lacked any equivalent mechanism. Progress in raising labour productivity was far too slow and rates of increase tended to fall until the mid-1980s. A number of factors were responsible, including an insensitivity to labour costs (a consequence of the structure of prices, wages and subsidies as they affected the operations of firms), the failure to innovate, the high degree of job security and uncertainties in the planning process (whereby firms found it necessary to hoard labour in case of late arrival of supplies, unforeseen accidents etc.). Employed personnel were used inefficiently, absenteeism was common and there was very high labour turnover. In general, it can be said that Soviet employees found few incentives to work hard or to stay on the job.

Many sectors of the Soviet economy suffered from labour shortages, especially the underpaid service sector and inefficient agriculture. Labour was also difficult to attract to underdeveloped or environmentally undesirable regions. These included the countryside (except in Central Asia with its rapid population growth) and much of the north, Siberia and the Far East. Needless to say, the difficulties of recruiting labour in the latter regions only added to the expense of resource development. Labour shortages, then, were an additional brake on Soviet economic performance.

Finally, the command economy suffered from serious problems of *pollution and environmental deterioration*, undermining the health of the population and having other economic effects. In view of the ramifications of these problems and their continuing importance today, an extended survey of them will be given in chapter 6.

Both sectoral and geographical problems therefore contributed to the growing difficulties of the Soviet command economy, although other factors, such as an ageing party leadership, incapable of facing up to the need for change, and various mistakes in economic policy, helped to precipitate an economic crisis over the late 1970s and early 1980s. This set the scene for the reforms of the Gorbachev period.

A Geography of Neo-Stalinism

Before we go on to consider the reforms of the post-1985 period which led to the demise of the command economy, it is worthwhile to summarize what has been said about that economy by considering its spatial structure. The reader will recall the discussion towards the end of chapter 2, where we argued that Soviet society is probably best thought of as 'state socialist' rather than 'capitalist'. Despite the arguments put forward by some geographers to the effect that, because state socialism produced deep-seated inequalities like capitalism, it 'does not exhibit any really distinctive dimensions' (see Knox and Agnew, 1989, p. 160), we would argue that it did in fact have a distinctive spatial structure. We can list the essential features of that spatial structure, of the 'geography of neo-Stalinism', as follows:

1 A well developed core–periphery structure, reflecting marked differences in levels of economic development and living standards. This is in part the product of a tendency towards 'incrementalism' – seeking to gain economies by allocating a considerable proportion of resources to those regions which have benefited most from previous investment decisions (Knox and Agnew, 1989, p. 162). The most obvious discrepancy in the former USSR was between the well developed European region and much of Siberia, the Far East, Central Asia and

the north, although there was considerable spatial unevenness even in the European territory. Despite an alleged commitment to regional equality and policies to spread industrial development for demographic, strategic and resource-related reasons, movement to ameliorate the centre–periphery differences was limited and there was little in the way of a coordinated spatial strategy.

2 The inbuilt conservatism of the system and the bias towards heavy industry ensured the continuing importance of traditional industrial regions with 'smokestack' industries, such as the Donets–Dnepr region of eastern Ukraine and the Urals. Relatively few regions felt the full impact of the new technologies which have had such repercussions elsewhere in the industrial world.

3 'Extensive' (i.e. resource-demanding) rather than 'intensive' (resource-saving) development, leading to waste of resources and environmental deterioration in the core, growing dependence of the core on the resources of the periphery and pressure to develop the latter in the cheapest and often most short-sighted manner.

4 Administration of the economy by sectors and tendencies towards 'narrow departmentalism' led to the development of a series of ministerial 'empires', lacking interlinkages, reducing the scope for scale economies, encouraging excessive transportation and leading to the economic overspecialization of many cities and regions, especially peripheral ones. Even regions with a broad sectoral structure, such as the Moscow city region, often showed evidence of inadequate interlinkages between sectors (*Moskovskiy stolichnyy region*, 1988, pp. 84–90).

5 The relative neglect of agriculture, transportation, consumer welfare and numerous services disadvantaged many rural regions and restricted the comprehensive development of cities, especially in the periphery.

6 A well developed hierarchy of well-being in the settlement structure whereby, in general terms, the best endowed settlements were the biggest ones with major administrative and political functions, and conditions deteriorated as settlements became smaller. This situation was encouraged by such factors as the tendency for political and economic actors to favour the bigger cities, restrictions on inmigration to big cities and the relative neglect of rural, agricultural and small town development. Unlike in many Third World countries, conditions in the big cities were to some degree protected by restrictive inmigration policies. This gives rise to the thesis of underurbanization, which argues that, because of such policies, the former USSR was less urbanized than would be expected given its level of industrial development, the mirror image of the situation in parts of the Third World

(Ofer, 1976). The overall picture regarding the settlement hierarchy varied according to the region in which the settlement was located (core or periphery) and by ministerial behaviour (some ministries did better by 'their' settlements than others).

7 The development of regional economies was greatly influenced by the 'military–industrial complex' with the progress of individual cities, groups of cities and even entire regions (including peripheral ones) very much bound up with the needs of the military machine.

8 Continental and inward-looking development induced by the long-standing tendency towards economic autarky. Isolation from the world economy and strategic necessity encouraged the development of some peripheral regions even when this was not entirely justified on economic grounds. Only from the 1960s were autarkic tendencies modified, encouraging further economic development along land frontiers, on coasts and at ports.

The Stalin and post-Stalin eras, therefore, produced a distinctive geography which was very much implicated in the growing problems of the command economy. It was those problems which Mikhail Gorbachev was expected to solve upon becoming Communist Party leader in March 1985.

The Economic Reforms of Gorbachev and Yeltsin (after 1985)

The political side of Gorbachev's reforms have been discussed in chapter 3. Here we are concerned with his economic policies.

As we have seen, Gorbachev had no blueprint for radical reform of the command economy and at first adopted a rather conservative approach. When at length it was decided to resort to more far reaching policies, these moved in the direction of decentralization, away from the neo-Stalinist command economy and towards the free market. This was in line not only with previous attempts at economic reform, such as the Brezhnev–Kosygin reforms introduced in 1965, but also with a broad trend in capitalist economies, which since the late 1960s had been moving away from Keynsian policies of demand management by the state (Hettne, 1990, pp. 57–8; Maley, 1995, pp. 50–2). Particularly significant was the Law on the State Enterprise (June 1987), which was designed to move enterprises on to full cost accounting over a two-year period and began the process of rolling back the frontiers of the state by starting the dismantling of the huge planning bureaucracy. The Law on Cooperatives (May 1988) removed the restrictions on cooperative economic activity which had been imposed in 1929 and opened the door to privatization. Other economic reforms of the Gorbachev period affected agriculture, foreign trade, land, property, financial affairs and other matters (for further details see Miller, 1993; White, 1993).

It is impossible to separate economic from political reform, especially in an economy where the economic had always been subservient to the political. The political turmoil of the late Gorbachev period has been discussed in chapter 3, where it was noted that this inevitably undermined the economic reforms. However, it is worth adding to that analysis by suggesting one or two reasons why the reforms proved inadequate on the economic level. The basic problem lay in trying to change an economy which had been constructed for one purpose (viz to serve the interests of the state, including its quest for military security) into one constructed for a very different one (viz to serve the market). Thus enterprises which had always operated in a non-competitive environment, assured of both their suppliers and their customers and only obliged to fulfil their state-imposed plan, now found themselves expected to meet their own costs and to find both suppliers and customers for themselves. A breakdown in inter-firm linkages was almost inevitable in these circumstances. Where firms had a virtual monopoly in a particular type of production, which was frequently the case and which had a certain logic to it under the command economy, they found that they could meet their profit targets by raising their prices and cutting down on output, thus making life easier for themselves. Unfortunately, this kind of thing had a knock on effect throughout the economy as firms lost their customers and/or suppliers and were unable to meet their debts. In a similar way, in an economy which was only partially marketized and where many rigidities and obstacles stood in the way of the operations of supply and demand, a firm changing its suppliers or its product mix, for whatever reason, caused problems for others. Inter-firm linkages and flows of goods to the consumer were also disrupted by local, regional and republican authorities, who began to try to prevent the outflow of scarce goods from their areas and thus to protect themselves from the growing economic chaos. Problems were also caused by inefficient transport, by strikes and by criminal activity. Government policy proved unhelpful, pursuing a destabilizing investment policy and making various other mistakes in monetary and fiscal matters which helped to fuel inflation. The government also found it ever more difficult to balance its budget. Thus, while quite efficient in demolishing the command economy, Gorbachev's policies had clearly failed to put a viable alternative in its place. The consequences in national economic performance are illustrated in table 4.3.

One of the problems which beset Gorbachev was his hesitation about

Table 4.3   Aggregate Soviet economic performance during the Gorbachev years

	1986	1987	1988	1989	1990
GNP	2.5	2.5	2.8	−0.1	−3.1

Source: Gregory and Stuart (1994, p. 236).

how far to proceed along the path of reform. His hesitation was no doubt linked to his beliefs as a communist. Even at the end of the Gorbachev era, many elements of the command economy were still evident, including state subsidy of many enterprises, controlled prices, a centralized supply system and much else. With the break-up of the Soviet Union, however, President Yeltsin of Russia felt free to pursue a policy of radical reform. The strategy adopted was that of 'shock therapy', a sudden and rapid move towards the market. The major argument for this approach to reform was that, although it would cause hardship in the immediate term, such problems would be short-lived and less severe than those which would ensue from a gradualist policy. It was also believed that a radical reform would make it less easy for the reform's opponents to mount an effective resistance. In January 1992, therefore, prices were freed, government subsidies to industry cut back and full autonomy and financial accountability extended to most state-owned enterprises. It was realized that the sudden withdrawal of government subsidies might lead to enterprise closure and mass unemployment, and therefore there was a commitment to a programme of social welfare to help the needy. At the same time, however, it was clear that the market would not function properly in conditions of monetary instability, encouraged by government overspending, and therefore the Yeltsin government pursued a policy of trying to balance the budget and achieve monetary stability. Finally, it was believed that opening up the economy to the world market would promote competition, inward investment and efficiency.

The results of these policies were not entirely as anticipated. Because of the high degree of monopoly in the Russian economy and barriers to the introduction of competition (including a non-convertible ruble, which discouraged imports), prices rose much more than expected, thus causing considerable hardship and economic uncertainty. The problems were probably exacerbated by the reluctance of Western governments to extend financial help. Moreover, the Yeltsin government found it impossible to maintain fiscal discipline and to reduce subsidies as much as had been hoped. This was partly because of fear of the consequences of inflation and mass unemployment, bolstered by the political struggle with parliament and various vested interests. There were also numerous institutional difficulties, including problems of controlling the central bank and of curbing the ruble-spending propensities of the other post-Soviet republics before they introduced their own currencies. The partial breakdown of economic relations between the newly independent republics (in what had previously been the single economic space of the USSR) only added to the problems. The widespread resort to bartering, not merely between republics but even between regions and individual firms, only partially alleviated the situation.

One of the problems which has beset marketization in Russia has been the lack of an appropriate infrastructure, such as banks and financial institutions. New legislation was therefore urgently needed to replace redundant laws appropriate only to the command economy. New laws on such things as property, land, privatization, taxation, banking, foreign investment, currency, social security, commercial practice and monopolies had to be passed, and this is an ongoing process. In addition, the training of a whole army of people equipped to operate in the new commercial environment was an obvious necessity. Needless to say, marketization could only be hindered by the political instability which afflicted Russia after 1991.

Perhaps the most far-reaching change promoted by the Yeltsin government was that of privatization. As we have seen, this began during the Gorbachev era with the introduction of cooperatives and family businesses, and it gradually extended to privatization in the realm of services and the retail sector, the establishment of joint ventures with foreign firms and the introduction of privatization to state-owned housing and to farming. Under Yeltsin all this was taken very much further, with the most important initiative being the extension of privatization to the industrial sector. The arguments for privatization are several (Fortescue, 1995). First, it is argued that privatization will result in firms being owned by people with a much greater stake in their success than was the case with the old industrial ministries, with consequent improvements in efficiency, access to investment capital and profitability. Second, it should result in the restructuring of industry, moving away from the traditional and inflexible reliance on large enterprises, monopoly production and inadequate cross-sectoral linkages. Third, proponents of privatization maintain that it should create a class of owners with vested interests in the market system, making it more difficult for the reform to be reversed in the future.

Several models of privatization exist. The most obvious one, of selling industry to the highest bidder, was unlikely to be successful in Russia, where there was an absence of private capital or of a tradition of entrepreneurialism. Some privatization along these lines took place before the Yeltsin government decided from October 1992 to embark upon two alternative models of privatization. These were, first, privatization by the management and employees of firms, and, second, privatizing by distributing vouchers to the population, which represented their nominal share in state-owned industry. The vouchers could then be used to buy shares, or simply sold for cash. Thus, by the end of 1993, more than two-thirds of the 14,500 large enterprises designated for privatization had been transformed into joint stock companies, mainly owned by their workers. It is estimated that about 60 per cent of industrial assets had been transferred to private hands by early 1995 (in the same year 73 per cent of all industrial enterprises were private, though more than two-thirds of all production came from enterprises in mixed ownership). The government meanwhile continued to control certain key sectors.

Some doubts must exist as to whether worker privatization is a form likely to produce the efficiency gains hoped for. There are also uncertainties about the most appropriate company structures for the new situation (for example, whether a predominance of small firms or large corporative structures). Current tendencies towards inter-firm alliances and agreements, while having potential advantages, have raised the fear that these may be attempts to reconstruct parts of the old Soviet industrial system with a dependency on government orders and subsidies (Fortescue, 1995, pp. 91–7). In the same way, there are claims that too much privatization has benefited the former officials and functionaries of the communist era, who have been able to use their positions to acquire shares, not always in an open and honest manner. In the Soviet period, high officials of party and state were known as the *nomenklatura*: hence the term *nomenklatura* capitalism to describe this phenomenon (Aslund, 1996). Needless to say, *nomenklatura* capitalism, with its overtones of political and even criminal activity in some cases, is hardly regarded as conducive to the workings of the free market. Optimists, however, regard this as a temporary phase in the transition to capitalism.

The Economy in Transition

What have been the effects of economic policy for the actual structure and functioning of Russia's economy? To answer this question it might be useful to begin by considering what might happen to the economy's structure at the theoretical level. It will be recalled that the Soviet economy was characterized by a strong orientation towards heavy industry, to what economists call the output of producer goods. This was partly for ideological and political reasons: in the early days, heavy industry was regarded as the prerequisite for transforming the USSR into a major industrial and political power, and for ensuring the numerical growth of the industrial proletariat who were the principal political supporters of the communists. For many years Soviet ideology taught the inherent superiority of heavy industry over consumer-oriented industry in ensuring rapid economic growth into the future and ultimate victory in the competition with capitalism. There were, however, other reasons: the inherent conservatism built into the Soviet system has already been noted. The vested interests of the big industrial ministries and the 'military–industrial complex' ensured that the Soviet system, once set on course by Stalin, was not to deviate very far from its preordained path without a titanic struggle.

Towards the end of chapter 2, it was noted that from the end of the 1960s the Soviet economy began to deviate more and more from those of the major capitalist countries. One of the effects of the transition from 'organized' to 'disorganized' capitalism was to reorientate the principal Western economies away from the traditional extractive and manufacturing activities and towards services. According to Treyvish et al., the USSR and its satellites moved only very slowly in this direction (Treyvish et al., 1993). While perhaps not necessarily wishing to copy Western experience in every particular, any conceivable modernizing reform in the Russian context must aim to move its economy away from the overconcentration on traditional 'smokestack' industries and towards the more advanced and sophisticated types of manufacturing and service activities.

Primary industries like mining have certainly experienced relative decline in the advanced capitalist economies and must be expected to do so in Russia. This would probably mean, for example, the reduced importance of coal mining, at least in the more expensive locations. It has already been noted that Soviet industry was long starved of capital for reconstruction, a problem which was exacerbated by the prevailing 'extensive' model of economic growth leading to a reluctance to retire antiquated equipment. Capital starvation has afflicted extractive as much as other kinds of industry. Russia's ability to benefit from its undoubted riches in energy, minerals, timber and other resources (and also from its relatively cheap labour) must depend upon the availability of capital investment and keeping costs down.

Much the same might be said about manufacturing industry. The problem here is that too much output has been oriented towards other kinds of economic activity which reflect the priorities of the command economy rather than of the market. Russia's long isolation from the world market meant that its resource requirements and often its costs have been too high and the quality of its output too low. It has also relied too heavily on traditional types of production and too little on high-technology manufactures. Since so much Russian manufacturing was geared up to the requirements of the 'military–industrial complex', a massive reorientation is required if Russian industry is going to be able to meet consumer needs at home and to compete on the world market. In the meantime, even those consumer industries which do exist run the risk of losing their markets to cheap and higher quality foreign imports. Similar problems face agriculture and the food industry, where there is not only a problem of capital shortages but also the demoralization and social disintegration produced by the disastrous Soviet policies towards the rural sector (see chapter 7). Altogether, deindustrialization seems an inevitable consequence of economic reform, possibly to an even greater extent than that experienced since the 1960s by the advanced capitalist economies.

The lack of a highly developed consumer sector in Soviet society meant that the Soviet citizen lacked the retail and consumer services which tend to be taken for granted by the citizens of advanced capitalist countries. In fact, the Soviet citizen lived in a world of shortages, spending hours on the simplest consumer tasks and often seeking in vain for the goods and services grudgingly supplied by the Soviet state. The new Russian economy can be expected to reflect consumer requirements more faithfully, even if the extent of the consumer sector's development very much depends upon how successful the market is in producing the wealth the citizenry expects. In a rather similar way, since the command economy provided many of its own service requirements or had no need for many services normal to capitalism, the stage may now be set for a big expansion of producer services.

Most commentators expected a period of disruption and economic decline to follow the introduction of economic reforms by the Yeltsin government in January 1992. In the event, the problems were even greater than anticipated. Some of the reasons for this have been alluded to above. Russia, being by far the biggest of the post-Soviet republics, responsible for more than 60 per cent of Soviet manufacturing output in 1985 (despite having only 51 per cent of the population), was probably best able among the republics to withstand the disruptions which came about as a result of the disintegration of the single Soviet economic space. Even so, the scale of the economic downturn came as a surprise. Since the Russian economy represents a major share of the old Soviet economy, its restructuring problems are commensurately great.

According to the official data, the fall in GDP over the period 1989–94 was about 49 per cent, with a slightly greater decline in industrial

Table 4.4   Changes in Russian GDP, and industrial and agricultural output, 1989–94 (in constant prices, 1989 = 100)

	1989	1993	1994
GDP	100	60.6	51.6
Industry	100	64.8	51.3
Agriculture	100	79.7	72.4

Source: Hanson (1996a, p. 20).

output (table 4.4). By any criteria, this is a massive production fall, and it continued during 1995 and 1996 (GDP fell by a further 4 per cent in 1995 and 5 per cent in 1996). It may be, however, that the figures are exaggerated: according to one estimate, the real GDP decline over 1989–94 is more like 30 per cent (Hanson, 1996a). Whereas under the old Soviet system there was an incentive for managers to inflate their production figures in order to demonstrate fulfilment of the plan, there is now an equally compelling incentive to minimize output figures in order to avoid taxes and other impositions. The official statistical service, *Goskomstat*, is unable to capture this effect. In fact, this serves to remind us of the inadequacy of the official statistics in measuring many aspects of the current economic and social transition. Not only has the statistical service itself suffered from cuts and problems in maintaining its activities, but there are also many unofficial or even small-scale official economic activities under way which escape its attention completely. Widespread tax evasion has severely hampered the government and indebtedness and fraud have spread throughout society.

Within a context of overall contraction (which continues as of 1998), the economy has experienced a remarkable restructuring. Table 4.5 suggests that this is very much in line with what might be expected given the points about restructuring made above. Thus the primary activities of agriculture and forestry have suffered contraction, relatively speaking, and so has industry, but wholesale and retail trade have grown relatively more important (and even increased their employment level). There has been a marked relative increase in service activity. Within the latter category, in fact, there has been differential change, with particular growth (both relative and absolute) in producer services. The overall picture is somewhat obscured by data problems, however. The employment data in table 4.5 suggest, for example, that the decline in agriculture's significance is more apparent than real. What has happened is

Table 4.5 Changes in the structure of Russian GDP and employment, 1989–94

	Output shares		Employment (1990 = 100)
	1989	1994	1994
Industry	38.9	30.4	78.5
Agriculture and forestry	16.2	6.8	92.0
Construction	11.3	9.1	80.0
Transport and communications	9.1	9.5	89.7
Retail and wholesale trade	8.8	16.6	103.4
Other[a]	20.1	27.6	91.3
All recorded employment			87.0

Note: [a] Includes public administration, public services, and marketed services (other than transport and distribution).
Source: Hanson (1996a, p. 20).

that, while output has suffered a lower decline than in industry, agricultural prices have failed to rise as much as industrial ones.

Table 4.6 (based on physical rather than value output indicators) shows the effects of restructuring within the industrial sector. The restructuring which has occurred suggests that while it was the disruptions associated with the break-up of the USSR and other 'supply side' factors which helped to account for the initial industrial decline, differential demand has become more important recently. Thus, over the 1990–5 period, fuels and electricity did better than most other subsectors, particularly engineering and light industries (clothing, shoes, textiles). The decline in engineering seems to reflect the overall decline in investment across the economy. It may also be a reflection of the reduction in military expenditure (assuming that the statistics include military expenditure). Cotton textiles initially faced disruption in their Central Asian supplies, once the Central Asian states were no longer part of the USSR, but more recently there has been competition from cheap foreign imports and possibly changes in household expenditure as a result of inflation. The latter factors have also affected other light industries. Food production may have suffered less because food is more central to household expenditure, but there is obviously the threat of more cheap food imports in the future. Fuels have benefited from export orders. The reason for the modest decline in electricity output is more debatable.

Industrial restructuring has therefore meant deindustrialization. While this is in line with the experience of other advanced economies, a

Table 4.6 Indices of Russian industrial output, 1992–5 (1990 = 100)

	1992	1993	1994	1995	1995 as % of 1994
All industry	75	65	51	50	97
Extractive	85	77	69	69	99.4
Manufacture	74	63	48	46	96
Electricity	96	91	83	80	97
Fuels	87	77	69	68	98
Ferrous metallurgy	77	65	53	58	109
Chemicals	73	58	44	47	108
Engineering	77	65	45	40	90
Timber	78	63	44	42	96
Building materials	78	65	47	45	95
Light industry	64	49	26	19	70
Food industry	76	69	57	52	92

Source: *Rossiya v tsifrakh* (1996, p. 281).

worrying aspect is the lack of any marked swing to high-technology production. The decline in engineering seems to reflect its general uncompetitiveness on the world market (with certain exceptions, like some armaments production), while the Russian consumer seems to prefer imported durables. Overall, then, there has been a swing against the higher branches of production and towards the more basic production of fuels and raw materials. This has been aided by a sharp fall in investment, a worrying development which no doubt reflects continuing political and economic uncertainties. If prolonged, this could militate against the development of capital-intensive production. Hanson argues that the restructuring which has occurred has resulted from net flows of labour rather than of capital (Hanson, 1996a).

The economic problems associated with transition have borne very severely on the mass of the population: hence the poor showing of democratic and market-oriented candidates in elections to the State Duma in December 1993 and 1995. Unemployment is on the increase. While this stood officially at 2.1 million in August 1995, using the methodology of the International Labour Organisation (ILO) this translates to 5.7 million or 7.8 per cent of the workforce. By the end of September 1997, according to ILO methodology, unemployment had risen to 6.5 million or 9 per cent nationally. This does not include the several million on short-time working or suffering enforced leave. Yet government subsidies and the traditionally paternalistic relationship between industry and employees have ameliorated the worst ravages of unemployment thus far. Workers have therefore been kept on payrolls, enjoying some health and social security as a result, even when monetary compensation has been minimal. Pensioners, government employees and others have suffered from the delayed payment of pensions and wages. The holding of multiple jobs, unofficial or even illegal commercial dealings, growing one's own food and similar activities have proved essential to the subsistence of many. Even then, some have been unable to make ends meet and there has been a sharp increase in homelessness, vagrancy and similar problems in many areas. Although the statistics are unreliable, it has been estimated that only about 10 per cent of the population experienced real increases in their income between 1991 and 1996. The rest have lost out (Morvant and Rutland, 1996).

The effects of economic transition will vary through space according to the local opportunities for economic activity, the inheritance from the past, local government policy and many other factors. This is the subject of the next chapter.

Five

The Changing Space Economy

The spatial implications of transition from a command to a market economy are bound to be profound. In terms of the theoretical discussion about the nature of Soviet communism at the end of chapter 2, it means a move from what we have called 'state socialism' to capitalism (whether or not this takes on a peculiar Russian form), with the integration of Russia and the other post-Soviet states into the global economy. The spatial effects of economic transformation will greatly depend upon the niche, or perhaps more accurately niches, which the post-Soviet states come to occupy within that economy. Because the entire situation is a dynamic one, the outcome is bound to be uncertain.

Marketization is inevitably changing the 'geography of neo-Stalinism' outlined in the previous chapter. But precisely in what ways the space economy will change is still a matter for much guesswork. As Doreen Massey has written of capitalist economies: 'the geography of industry is an object of struggle. The world is not simply the product of capitalism's requirements' (Massey, 1984, pp. 7–8). The changing space economy of Russia and its neighbours will reflect the outcome of ongoing political struggles, both national and local. In a sense, this was always the case even in the Soviet era, as Soviet policy reflected the outcome of behind-the-scenes struggles among politicians and between them and organizations, groups and other interests (Skilling and Griffiths, 1971). In the post-Soviet situation, these struggles have often become more open and less amenable to direct political control. They will influence, and be influenced by, the movement of capital, both international and homegrown. Our uncertainty about the outcome of all this is compounded by the way in which capitalism's requirements themselves change through time as new technologies and forms of industrial and economic organization emerge. Massey argues that the spatial effects of this dynamic situation are highly variable, with different regions and places experiencing different outcomes as the 'presently emerging spatial division of labour' is superimposed upon the social and economic structures left by earlier ones (Massey, 1984).

The economic and social structures and the associated spatial patterns which were the product of the Soviet mode of development have been briefly discussed in previous chapters. It was asserted, for example, that Soviet economic development was characterized by considerable spatial inequality, despite a Soviet commitment from the earliest days to evening out such spatial discrepancies. These inequalities have been studied by numerous Western and Russian scholars (for example, Bahry, 1987; Liebowitz, 1987, 1991; Schiffer, 1989; Dmitrieva, 1996), who tend to agree about the resilience of such patterns even if they disagree about either their underlying causes or their propensities to change. The disagreements are fuelled by the many data-related problems which such studies involve. The most notable spatial differences which have been identified, embracing both levels of economic development and living standards, have been those between the more developed 'northern' Union republics of the former USSR, namely the Russian Federation, the Baltic republics, Belarus' and Ukraine, and the southern tier, consisting of Moldova, the Transcaucasian republics, Kazakhstan and Central Asia. Scholars have also traced notable dimensions of inequality within the Russian Federation. These were most obvious along an east–west axis, with Siberia and the Far East suffering numerous disadvantages compared to the more developed and better populated regions west of the Urals. Dmitrieva's recent study supports the latter conclusions with respect to living standards, but her findings indicate a more marked north–south gradient in economic development levels, both within the Russian Federation and in the former USSR as a whole.

Further insight into the spatial patterns bequeathed by the Soviet era is given by studies of the variant economic structures of the different regions of the country. Nefedova and Treyvish, for example, describe an attempt to classify the regions of European USSR according to their occupational structures at the end of the 1980s (Nefedova and Treyvish, 1994, pp. 10–11) (see figure 5.1, showing Russian territory only). Although the basis of the classification is not described in detail, the study suggests the wide differences between regions in occupational structure and, by implication, in their degree of 'modernization'. Thus only three regions – the 'capital' regions of Moscow, St Petersburg (Leningrad) and Kiev – are categorized as 'young, post-industrial', reflecting their relatively well developed service sectors by Soviet standards (even if such regions would not merit this designation in the West). At the other end of the scale, a number of regions in the North Caucasus, bordering on Ukraine and in central and western Ukraine itself are described as pre-industrial, or in transition to industrialization. The main shortcoming

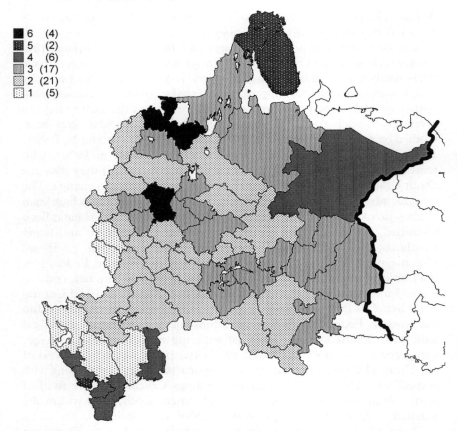

Figure 5.1   A typology of regions in European Russia in the late 1980s according to occupational structures.
Source: after Nefedova and Treyvish (1994, p. 11).

with this classification lies in its apparent assumption of a normative model of industrial development, but it does at least reflect the considerable regional economic differences existing at the end of the Soviet period.

The same authors describe a classification of industrial regions for the whole of the former USSR, based upon their economic profile, their degree of development and their future prospects (Nefedova and Treyvish, 1994, pp. 11–12). Notable here is their depiction of the Urals, certain regions of Ukraine and to some degree the Central Region in Russia as regions without strong economic prospects, though their problems were largely masked in the Soviet era by the peculiarities of the command economy.

Another author, Bylov, has attempted to classify Russian regions into four types according to the major social problems facing them in the transition period (Bylov, 1995):

1 Big industrial centres with highly developed economies (for example, Centre, Urals, Northwest). These are regions characterized by high levels of existing or potential unemployment and in need of industrial restructuring.
2 Big agricultural regions, mainly in southern Russia, now facing large influxes of migrants and refugees and with limited present-day employment opportunities.
3 Resource-extracting regions of northern and eastern Russia, especially those developing in the relatively recent past, which now face depopulation as a result of ecological problems and a poor quality of life.
4 Economically backward national territories (especially in the North Caucasus and Siberia) with low levels of economic development, a surplus of young people, a relatively unskilled labour force and high levels of unemployment. The problems of these regions are exacerbated by ethnic tensions and conflicts.

A general indication of the geography of employment in industry and agriculture in the mid-1990s can be found in figures 5.2 and 5.3.

Regional Economic Change since 1991

How precisely has economic change since 1991 altered the 'geography of neo-Stalinism' discussed previously? This is obviously a very complex question which we cannot even begin to address without an adequate body of statistics. Unfortunately, the available statistics, though in many ways better than in the secretive Soviet era, still leave a great deal to be desired. One reason for this derives from the very political and economic changes which are the subjects of this book. As noted already, the official statistical service, *Goskomstat*, has not been immune to the many problems which have afflicted Russian society in recent years, among which a shortage of funds has been particularly difficult. Local statistics are especially hard to come by, which is a pity, since it is at the local level that some of the most interesting changes are taking place. Statistics

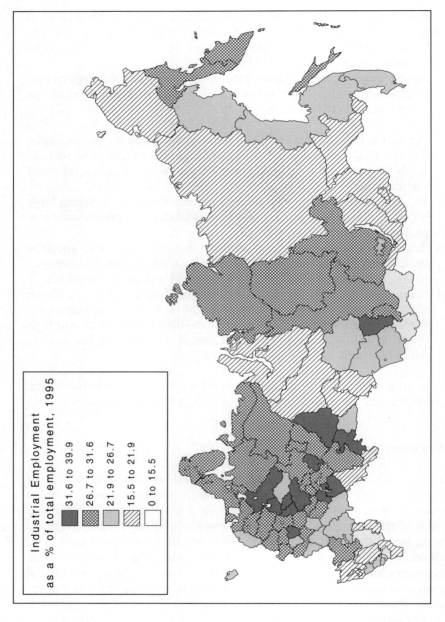

Figure 5.2 Russia: regional distribution of industrial employment, 1995.

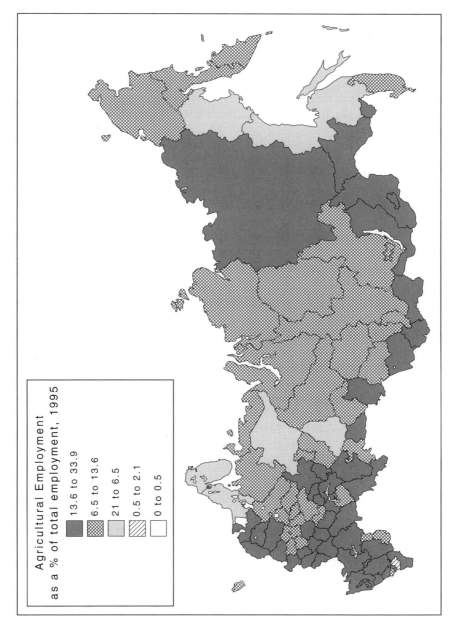

Figure 5.3 Russia: regional distribution of agricultural employment, 1995.

Agricultural Employment
as a % of total employment, 1995

- 13.6 to 33.9
- 6.5 to 13.6
- 21 to 6.5
- 0.5 to 2.1
- 0 to 0.5

which are averaged at the level of the oblast will miss out on important local variations, such as the situation in individual cities or districts. Statistics showing changes in sectoral output and employment are also inadequate. Another serious problem is that official statistics fail to include material on unofficial activity and the 'black' economy. This not only makes it impossible to know the real level of economic activity, but makes many of the data on employment and unemployment virtually meaningless.

Despite these problems, a number of studies have been undertaken to examine regional economic change since the end of the Soviet era (Nefedova and Treyvish, 1994; Bradshaw and Palacin, 1996; Bradshaw and Shaw, 1996; Sutherland and Hanson, 1996), and their main findings will be summarized below. Since regional economic change takes place on a number of dimensions, not all of which are clearly coordinated with one another, we will briefly discuss each kind of change individually before examining their aggregate effects at the end.

Production decline The national decline in industrial activity since 1989 was discussed in chapter 4, where it was also pointed out that the official statistics almost certainly exaggerate the true picture. Decline will naturally have differential regional effects, as will the economic restructuring which has gone on. The exact regional effects will vary according to the time period covered, but in general the available studies underline the particularly severe contraction which has occurred in the North Caucasus (especially in the republics), along parts of the frontier with the Baltic states, across parts of central European Russia and the Urals, and in certain regions of southern Siberia and the Far East (figure 5.4). The production decline in agriculture was less severe, but in general terms corresponded with that in industry. No region has escaped the effects of economic recession over this period.

Unemployment The geography of unemployment is particularly difficult to study, not only because of the prevalence of different definitions of unemployment (for example, those claiming unemployment benefit as against those actively seeking work), but also because of the widespread incidence of employees on short-time working, on enforced vacations or only nominally employed. Analysis of the real level of unemployment has to take into account a multiplicity of definitions, especially since there is still a marked reluctance on the part of many employers to dispense with labour even when there is little or no work for them (Rutland, 1996a). Employer attitudes may be influenced by the effects of economic transition and also by cultural factors, and may be presumed to vary spatially.

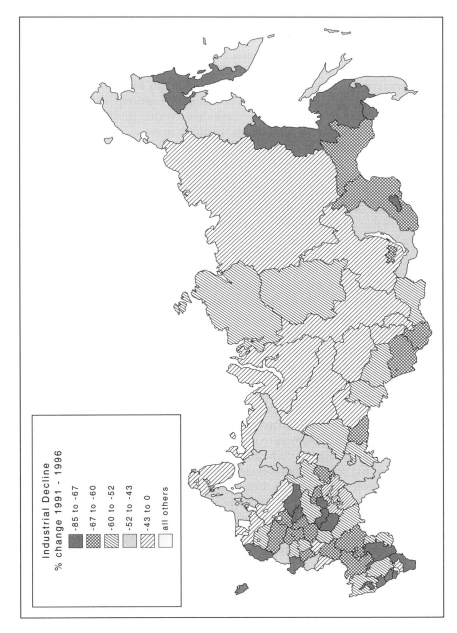

Figure 5.4 Russia: regional variations in industrial production decline, percentage change, 1991–6.

In general, unemployment is more of a problem in European Russia than it is in the east. Nefedova and Treyvish recognized three nodes of unemployment in 1994: an area across central European Russia, a belt in the northwest running northeastwards from the Estonian frontier towards the Arctic coast and the North Caucasian republics. In early 1996, the deepest pockets of unemployment were reported in Ingushetia in the North Caucasus (23.5 per cent, partly the result of refugee migrations from the war in Chechnya), Ivanovo in the Central Industrial Region (13 per cent, as a result of the problems of the textile industry) and Udmurtia in the Urals (9.4 per cent, reflecting problems in the defence industry).

Prices, incomes and living standards   A very distressing aspect of the economic transformation has been the huge inflation which began after the Yeltsin economic reforms in 1992. This inflation has been differentiated across the territory, so that prices in some regions of the country are far higher than in others to an extent unparalleled in western Europe. Thus in 1996, a basic food basket cost some 68 per cent of the national average in Ul'yanovsk Oblast on the Volga, but 185 per cent in Yamalo-Nenets Autonomous Okrug in northwest Siberia. Such discrepancies have been caused by the development of regional markets, partly fostered by inadequate transportation and supply systems, and partly by the activities of regional and local government. The cost of living in the mid-1990s tended to be high in Moscow, in the north of European Russia and across much of Siberia and the Far East, especially in the extreme north and east. These costs were partially compensated by differences in incomes, however, reflecting the geography of living costs but also the effects of marketization on the various economic sectors. Again these differentials are far higher than in western Europe. In the mid-1990s, incomes were higher than average in Moscow and across the resource-producing regions of the north, but low across south-central and south European Russia and in a belt crossing southern Siberia. Finally, combining data for incomes and costs of living, one arrives at a picture of the geography of living standards across Russia (figure 5.5). Unfortunately, in the high income areas, incomes are frequently cancelled out by high prices, whereas the opposite benefit is less apparent in low income areas. Altogether, in the mid-1990s, relatively high living standards were apparent in Moscow, to some degree in St Petersburg and in one or two regions of the north and Siberia. South European Russia and much of southern Siberia had low standards. It is probable, however, that incomes in many southern regions were boosted by informal activities like gardening,which are not reflected in the official statistics.

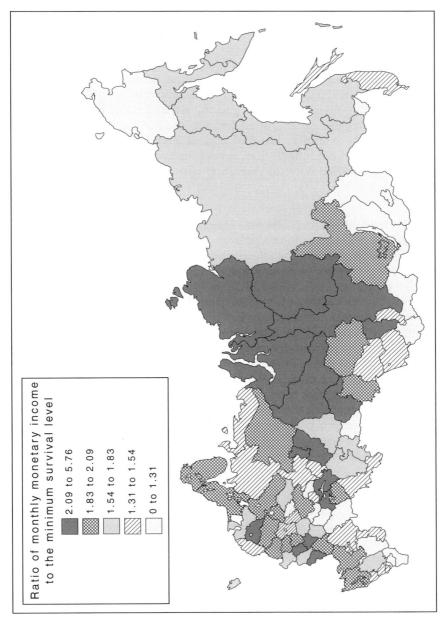

Figure 5.5 A geography of living standards in Russia, 1996 (regional variations in the monthly monetary income/minimum survival ratio, percentage of Russian average).

Ratio of monthly monetary income
to the minimum survival level

- 2.09 to 5.76
- 1.83 to 2.09
- 1.54 to 1.83
- 1.31 to 1.54
- 0 to 1.31

Interregional migration   Migration has naturally been seen as an important indicator of regional well-being or stress, although reflecting political as well as economic influences. Russia has experienced significant interregional migration flows since the break-up of the USSR in 1991. A map of the migration balance in the early to mid-1990s is given in figure 5.6. Negative migration balances were recorded across much of the Northern economic region, northern Siberia and the Far East. The highest outmigration was recorded in the far north and extreme east. By contrast, most of central and southern European Russia experienced net inmigration, apart from certain re-publics of the North Caucasus. The issue of migration will be further examined in chapter 7.

Other indicators of regional economic change   In addition to the above dimensions of regional economic change, scholars have been interested in indicators which reflect the effectiveness of economic reform and the potential for a region's economic success in the future. Unfortunately, because a good deal of unofficial and small-scale private activity escapes the official statistics, reliance often has to be placed on indirect indicators of economic success. Nevertheless, by analysing the data on such diverse phenomena as hard currency inflows to a region, investment flows both domestic and foreign, rates of privatization and the proportion of regional fixed capital which is privately owned, commercial bank assets, activities of small enterprises, evidence for high-technology production and the incidence of joint ventures, tentative conclusions can be drawn about how far regions have progressed down the road of marketization and what are their prospects for the future. The effects of such factors will be considered later.

In conclusion to this section, it is worth considering how these various patterns of economic change appear to be interacting in the fortunes of particular regions. One attempt to do this is by Nefedova and Treyvish (1994, pp. 31–7). By statistically measuring the combined effects of three indicators (viz the dynamics of production, unemployment levels and migration balance), these authors identify the least favoured and the most favoured regions as they were in 1993–4 (figures 5.7 and 5.8). The least favoured regions are of three kinds:

1 'Stressed' regions of armed territorial conflict, where all economic indicators are negative (North Osetia, Ingushetia and Chechnya in the North Caucasus).
2 Impoverished peripheral regions suffering from increases in living costs and high outmigration, but not necessarily high unemployment.

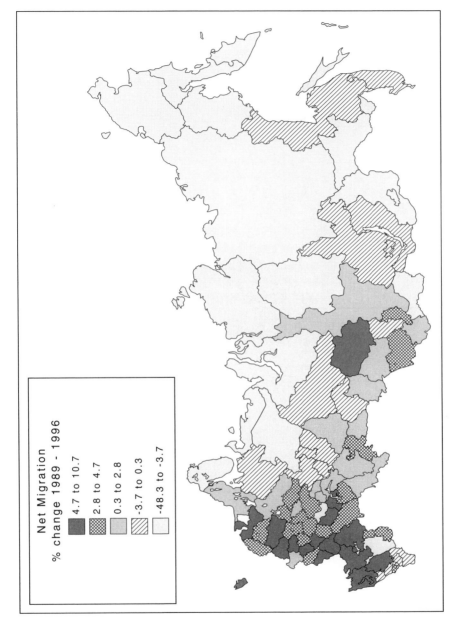

Figure 5.6 Russia: regional variations in population change due to migration, 1989–97.

Net Migration
% change 1989 - 1996

- 4.7 to 10.7
- 2.8 to 4.7
- 0.3 to 2.8
- -3.7 to 0.3
- -48.3 to -3.7

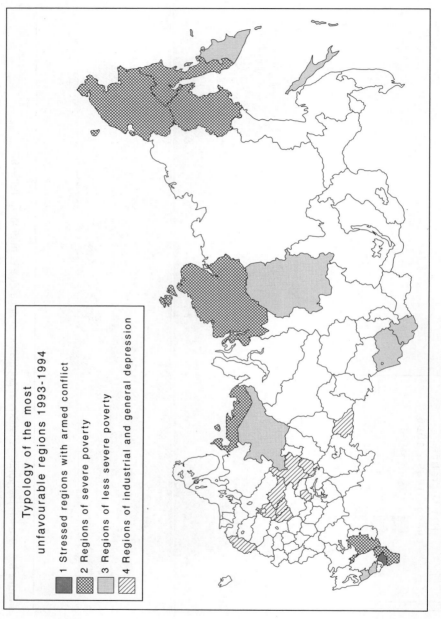

Figure 5.7 Russia: regional 'losers' in economic transition.
Source: after Nefedova and Treyvish (1994, p. 35).

Typology of the most
unfavourable regions 1993-1994

1 Stressed regions with armed conflict

2 Regions of severe poverty

3 Regions of less severe poverty

4 Regions of industrial and general depression

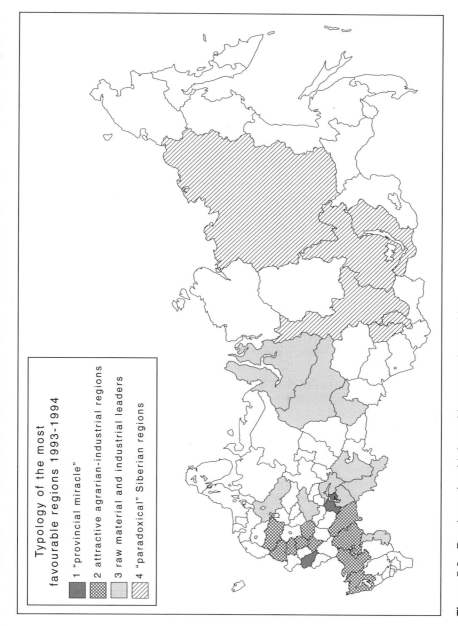

Typology of the most
favourable regions 1993-1994

1 "provincial miracle"
2 attractive agrarian-industrial regions
3 raw material and industrial leaders
4 "paradoxical" Siberian regions

Figure 5.8 Russia: regional 'winners' in economic transition.
Source: after Nefedova and Treyvish (1994, p. 36).

In the case of the North Caucasus, they are also suffering from severe production declines. They include both ethnic and non-ethnic peripheries, situated in parts of the European north, Siberia and the extremities of the Far East.

3 Regions of classical industrial decline, suffering from unemployment but also experiencing inmigration. These are primarily located in the Central Industrial Region.

The most favoured regions are of four kinds:

1 Regions which seem to be favoured by purely local factors, namely Ul'yanovsk on the Volga and Belgorod in the Central Industrial Region.
2 Agro-industrial regions to the west and south of Moscow and also in a belt stretching from Krasnodar in the North Caucasus northeastwards towards Saratov on the Volga. Although suffering from industrial decline, these regions have moderate unemployment and cost of living increases, with considerable inmigration.
3 Energy-producing and a few industrial regions with good resources, broad economic profiles and sometimes good responses to economic reform. They include Nizhniy Novgorod Oblast (Volga–Vyatka region), Samara Oblast (Volga region) and Tyumen' Oblast in West Siberia.
4 Some 'paradoxical' Siberian regions, mainly in the centre and northeast, which, though rapidly losing population, score positively on other indicators.

It is, of course, probable that these patterns will continue to change in the years ahead. What can be said in the meantime is that the spatial changes which have occurred since 1991 have so far done little or nothing to ameliorate the inequalities and dislocations inherited from the past. Indeed, in many respects they have made them worse.

Factors Influencing Regional Economic Change

What are the principal factors which appear to be influencing regional economic change in the transition to capitalism? Obviously this question is a particularly difficult one to answer at the present time. For one thing, the transition is still very much under way, the Russian economy has yet to recover from the deep recession which followed the destruction of the command economy and data on the processes now at work in that economy are still hard to come by. At the beginning of this chapter reference was made to the work of Doreen Massey, who warns of the uncertainties which beset the development of the capitalist economy. Geographers long ago moved away from trying to explain industrial location by using such simple constructs as Weber's least cost model or profit maximization models, or even by investigating the behavioural peculiarities of entrepreneurs and firms. Industrial location and regional economic change are now seen as processes which take place within the wider contexts of national, international and even global economic forces, the complex interplay of external and internal changes and decisions. Regional economic change in the Russian situation must likewise be seen as the product of forces both external and internal to Russia itself.

In a successful transition, regional economic change will consist of two processes: (a) the adaptation of existing economic activities to the new market economy (adaptation may also include the closure of such activities); (b) the development of new activities. In this context, the economic structure inherited from the past is clearly important. In a recent study, Sutherland and Hanson have analysed the concept of 'value subtracting' activities in the context of the old Soviet economy (Sutherland and Hanson, 1996). The concept draws attention to the fact that those activities whose final products are worth less in world market prices than the raw materials which were used in their manufacture are unlikely to survive in the new market conditions. This fairly obvious point is more pertinent to the Russian economy than it might seem at first sight, since the old Soviet system did manufacture many items which were perfectly usable in that system but which, because of low quality or dated technology, might actually be worth less on the world market than the inputs which they consumed. Sutherland and Hanson cite the work of other scholars who suggest that this might have been true of such sectors as light industries, food processing and some branches of engineering, while other activities such as extractive industries, metallurgy and other engineering branches are likely to be more successful. Clearly, then, those regions with more of the 'successful' economic sectors inherited from the Soviet era are likely, other things being equal, to do better than those with more of the value subtracting branches.

There is a good deal of evidence to support this conclusion. Thus the problems of the Central Industrial Region to the east of Moscow have been very much bound up with the difficulties of the light industrial branches, especially textiles, located there. The city of Ivanovo, once referred to as the 'Russian Manchester', is the prime case. Ivanovo's industries collapsed initially as a result of being cut off from their Central Asian suppliers but now because of competition from imports. The opposite state of affairs is found in Tyumen' Oblast and other parts of the north supplying natural resources to the world market. Here the problems of economic decline, though severe, have been less serious than elsewhere, and future prospects are relatively good. Similarly, as of 1996, the industries of the Urals (predominantly fuels, metallurgy, petrochemicals and chemicals) were suffering less severely than those of other regions, having been able to benefit to some degree from the liberalization of prices and export markets. In general, Sutherland and Hanson conclude that those economic activities are more likely to succeed in market conditions the less the processing that has to go into their products. They are thus rather sceptical of claims regarding the viability of Russian engineering, which requires considerable capital investment and restructuring if it is to succeed on the world market. In the Soviet period, of course, much of the engineering industry was bound up with the requirements of the military machine. With the reduction in government orders, and the competitiveness of the export markets for arms, the defence industries face particular difficulties, clearly affecting those regions where they are concentrated (notably the Northwest, the Central Industrial Region, Volga–Vyatka, the Urals and southern Siberia).

In their study, Sutherland and Hanson specifically test the importance of economic structure (as reflected in employment) in determining regional economic fortunes in 1992–3 (Sutherland and Hanson, 1996). Despite many problems in the data, they conclude that the role of inherited economic structure was only a modest one, and suggest that new patterns may now be at least as important as old ones in determining economic outcomes. What factors might lie behind these new patterns?

Standard location theory teaches us that, with the exception of a few resource-intensive activities, it is access to the market which is most important for the majority of economic activities today. Thus it is those regions with central or strategic locations, and with developed infrastructures, which are best placed to benefit from current opportunities. Remote locations, like the far north and Far East, have suffered from the withdrawal of government subsidies and rising transportation tariffs: hence the rising living costs and depopulation noted above. By contrast, existing transportation nodes like Moscow (which dominates the Russian railway network) and other regional centres, or ports like St Petersburg, Novorossiysk or Vladivostok, have clear advantages. Give the character of urbanization during the Soviet period, it is big cities which have the best developed infrastructure, the skilled labour forces, the educational and scientific research facilities and the access to markets which commerce and industry require. The tendency to nodality is reinforced by the poorly developed transportation systems in regions and rural districts. Many regional centres are set to benefit from the reorientation of supply systems (the development of wholesaling) and, in the short term at least, from the development of regional markets following the breakdown of the command economy. Agglomeration economies (i.e. the benefits of being close to other economic activities and of developing linkages) will benefit national and regional centres.

Little wonder, then, that, in addition to a number of resource-producing regions, it is large metropolitan centres like Moscow and St Petersburg as well as a number of regional centres which tend to lead the field in such matters as the number and size of banks, foreign direct investment, the establishment of joint ventures and private entrepreneurial activity. Such development can become self-reinforcing, whereby the availability of investment capital, for example, or the flow of skilled labour follows previous successful development (Filatochev and Bradshaw, 1995). Here the special case of Moscow must be mentioned. According to a methodology employed by Hanson, the city of Moscow exceeds all other regions on an aggregate indicator of attractiveness to foreign investors (second comes the energy-producing Tyumen' region and third St Petersburg) (Hanson, 1995). Quite apart from its geographical and infrastructural advantages, of course, Moscow has the unique merit of being Russia's capital. The economic advantages of being close to the source of political authority in Russia have a history which dates back to the days of the tsars.

A further factor which relates to geographical position is access to foreign markets. Clearly, the changes in Russian foreign trade which have followed the breakdown of the command economy (for more on this, see chapter 11) will benefit some regions more than others (see Kirkow, 1997). Proximity to ports, international airports, coasts and foreign borders can be a clear advantage in some circumstances. Some places in the Far East have begun to benefit from a growing local trade with China, for example, while other centres will benefit, or alternatively lose out, from the reorientation of trade between the post-Soviet states. The poor showing of Pskov Oblast on the border with Estonia is a warning of what can happen in the wake of a breakdown of linkages and a growth of international tensions.

Finally, one should mention the political factor behind economic change. Chapter 3 briefly discussed the division of powers between the Russian Federation government and its 89 'subjects' which came into being as a result of the 1993 constitution. From 1994, regional revenue sharing arrangements were to be within the framework of a formula-based grant mechanism. Thus some regions were to be net contributors to the federal budget, others net recipients, according to their defined needs. In reality, however, a fair amount of bargaining and discretion enters into the process, with some resource-rich republics (such as Tatarstan with its oil and Sakha with its diamonds) able to negotiate special arrangements and treaties. Clearly, such processes are likely to have significance for the economic futures of different regions.

Also significant, at least over the short and medium term, are the different policies pursued by regional and local governments themselves. One of the features of the reform process in Russia has been the much greater power over and responsibility for their own economic prospects which now rests with regional governments, compared to the situation in the Soviet period. The importance of this is reflected in the much cited contrast between Nizhniy Novgorod and Ul'yanovsk regions in the years after 1991 (OECD, 1995). Whereas the Nizhniy Novgorod authorities favoured policies of economic liberalization, privatization and attraction of inward investment, Ul'yanovsk's policies were based around price controls on food products (it is an agricultural area), restricting privatization and administrative controls over production and trade. Hardly surprisingly, Nizhniy Novgorod proved more successful in attracting foreign investment and in restructuring, whereas Ul'yanovsk was able to protect living standards, at least over the short term.

One source has classified Russian regions, in accordance with their attitudes towards economic reform, into extravert and introvert regions (OECD, 1995). The major extravert regions are: (a) resource-rich provinces in sparsely populated regions of northern European Russia and Siberia; and (b) major commercial centres and entry points, such as Moscow, St Petersburg, Arkhangel'sk, Astrakhan, Kaliningrad, Khabarovsk, Murmansk, Nakhodka, Rostov and Vladivostok. Introvert regions are: (a) provinces dominated by the military–industrial complex; and (b) agro-industrial regions with developed industry and self-sufficiency in food production, mainly in central and southern European Russia and southern Siberia. According to this source, extravert regions tend to take a liberal view of economic reform (at least as far as their authorities are concerned), agro-industrial regions favour protectionism and military–industrial regions vary in their responses.

Winners and Losers in Regional Economic Change

In summary, can any overall classification of regions be made in accordance with their inherent characteristics and their economic prospects? As indicated above, this whole process is made extremely difficult by the inadequacy of the statistics and also because of the newness of the transition process. What we offer below is a tentative classification of regions, based on the results of recent work (Hanson, 1995; Bradshaw, 1996), with some comments on their prospects.

Russia's regions can be divided into five types:

1 *Agricultural regions*. These embrace the agro-industrial regions noted above, mainly located in the southern part of European Russia and parts of southern Siberia. Outside the big cities, their infrastructures tend to be poor and their agricultural sectors have been resistant to change. Politically, these areas tend to be conservative, with a similar stance on economic reform. For this reason, these regions are sometimes referred to as the 'red belt'.

2 *Gateway and hub regions*. These include coastal, port-related and some frontier regions (notably, the city of St Petersburg, Krasnodar in the North Caucasus and Primorskiy Kray in the Far East), and Moscow as the most obvious commercial hub. They may also include some regional centres, like Sverdlovsk Oblast (with the city of Yekaterinburg) in the Urals and Novosibirsk in West Siberia. These regions enjoy many of the geographical and infrastructural advantages noted above and many of them have foreign trade links. They frequently take a liberal stance on economic reform and many are playing an increasingly important role in the development of services. A number of these regions are also well placed to play a role in the development of new, high-technology industries.

3 *Resource regions*. These are regions which came into prominence in the Soviet era producing the resources so badly needed by the command economy. Nowadays, with more control over their own affairs, they are exporting their resources on the world market. But many of them are suffering from the effects of remoteness, being largely situated in north European Russia and across Siberia and the Far East, from soaring transportation costs, the withdrawal of government subsidies and outmigration. Once the real costs of production and transportation are brought to bear, the continued profitability of all their resource-extracting activities cannot be guaranteed. This is especially the case given the dire need for capital investment. Equally, however, there may be the possibility of developing local

resource-processing activities in some instances. The reorganization of activity, with some pruning of unprofitable ventures, seems likely in the future.

4 *Old industrial regions.* These include many of the core industrial regions of the command economy, oriented towards producer goods, military output and the consumption of vast quantities of natural resources. They include many regions in the Central Industrial Region, Volga–Vyatka, the Volga region, the Urals and southern Siberia. The future of these 'smokestack' industries must frequently be dubious; such regions have often taken a conservative political and economic stance, favouring protection and government subsidy.

5 *'High-tech' industrial regions.* According to Hanson (1995), some industrial regions are characterized by the presence of a considerable number of research institutes and analogous facilities which were traditionally tied to the military–industrial complex. This considerably enhances their potential for economic revival in a market economy. Moscow, St Petersburg and Nizhniy Novgorod are obvious 'high-tech' centres, although these are classified as gateway and hub regions. Among the industrial regions, Voronezh in the Central Black Earth region, Saratov and Samara on the Volga and Chelyabinsk in the southern Urals are some that can usefully be classified as 'high-tech' regions.

Bradshaw (1996) mentions the possibility of the emergence of new growth centres and regions in Russia's burgeoning market economy, such as regional centres for the service industry (see below) or the development of a Russian 'sunbelt' in the south, close to the Black Sea coast. Looking further to the future, then, the industrial patterns inherited from the Soviet era may have diminishing relevance to Russian economic development. But everything depends on what policies are pursued nationally and locally.

Sectoral Developments: Energy, the Defence Sector, Services, Transportation

Now that spatial change in Russia's transitional economy has been considered in broad perspective, the final section of this chapter will discuss spatial and associated changes in four crucial sectors: energy, the defence industry, services and transportation.

The energy industry

The energy industry is crucial to Russia's economic transition. Russia is one of the world's most important energy producers, accounting for 13 per cent of production in 1995, second only to the United States with 20 per cent. The country produces 28 per cent of the world's natural gas, 12 per cent of its coal and 11 per cent of its oil. Energy production accounts for about 20 per cent of Russia's GDP.

Chapter 4 showed that the rise of the energy industry was a key element in Soviet industrialization policy, essential to the needs of the command economy and latterly both for supplying the USSR's allies in eastern Europe and for earning hard currency on the world market. In the post-Soviet era, energy production is still vital both to Russian industry and to those of its near neighbours, but its role in exports has become even more significant. Thus in 1995, exports of oil and gas earned Russia more than US$25 billion, or 40 per cent of its hard currency earnings. As Russian industry has collapsed, so the country has become even more dependent on energy.

Chapter 4 spoke of the rising problems of the energy industry in the latter part of the Soviet period as ever greater reliance had to be placed on the remote and expensive resources of the east (see figure 4.3). Unfortunately, these problems have continued into the post-Soviet era, when the sector's problems have been exacerbated by other difficulties, such as shortages of equipment and supplies. The industry has been unable to escape the problems which have afflicted all other areas of the Russian economy. Thus Russian oil production has been falling continuously since 1987, with a particularly sharp decline over the period 1990–2 (table 5.1). Likewise, coal production has declined since 1988 and gas declined from 1991 to 1995. While such factors as the increasing dislocation of the Soviet and post-Soviet economies played an important role in production decline initially, falling demand from failing industries has become more significant over the recent period.

The Russian oil industry is now paying the price for years of capital starvation, being still largely dependent on the infrastructure of the Brezhnev era (Sagers, 1996). In 1996, just under 70 per cent of Russian production came from the swamplands of West Siberia. Unfortunately, following the depletion of the unique Samotlor oilfield in this region as a result of Soviet shock development tactics, the remaining reserves (amounting to about three-quarters of Russia's proved total) are scattered about in small and remote fields which are going to be very expensive to develop. Other reserves are located in the Volga–Urals region (accounting for about a quarter of Russian production in 1996 but quickly declining), the north-east of European Russia (the Komi Republic and the Nenets Autonomous District), off Sakhalin in the Far East and in the North Caucasus. Western estimates suggest a national oil reserve of 50–60 billion barrels, or less than 10 per cent of the world total (official Soviet estimates were up to three times greater). Two-thirds of these reserves are in West Siberia. The oil industry therefore has consid-

Table 5.1   Russian fuel and energy production, 1970–96

	1970	1980	1990	1991	1992	1993	1994	1995	1996
Oil and gas condensate (million tons)	285	547	516	462	399	354	318	298	293
Natural gas (billion cubic metres)	83.3	254	641	643	641	618	607	595	601
Coal (million tons)	345	391	395	353	337	306	272	262	255
Electricity (billion kWh)	470	805	1,082	1,068	1,008	957	876	862	846
Thermal	373	622	797	780	715	663	601	583	
HEP	93.6	129	167	168	173	175	177	177	
Nuclear	3.5	54	118	120	120	119	97.8	99.5	

Source: *Rossiyskiy statisticheskiy yezhegodnik* (1996, pp. 511, 514, 516, 517), *Russian Economic Trends* (1997, no. 1, p. 102).

erable potential, though much less than was claimed by the Soviets, and it will require a great deal of investment in exploration, the drilling of new wells and the replacement or development of the infrastructure to take advantage of it. Other problems of the industry include government price controls (though prices are now rising), indebtedness of clients and the activities of vested interests who siphon off profits and deter foreign investors (Rutland, 1996b). The privatization of the industry took a peculiar form, establishing a dozen or so vertically integrated conglomerates operating in an oligopolistic environment. Foreign investment is clearly needed, but deterred among other things by political suspicion and managerial reluctance. At the moment foreign involvement is slowly increasing. The industry's future is therefore a rather uncertain one.

The natural gas industry appears to be much more promising, with Russia claiming some 35 per cent of the world's reserves. Almost 80 per cent of Russia's commercial grade gas reserves are in the remote northern coastal territories of West Siberia, which accounted for over 90 per cent of production in 1994–5 (Sagers, 1995a). *Gazprom*, the monopoly privatized producer, has been remarkably successful in sustaining overall production throughout the most difficult period of Russian economic decline. In 1995, about 12 per cent of production went to other CIS members and about 20 per cent to foreign exports. There are hopes that production will increase to the end of the twentieth century and into the next. However, although this seems possible from the viewpoint of reserves, much depends upon the performance of the domestic economy as well as on foreign markets. The major problems for the foreseeable future, therefore, stem mainly from the demand rather than the supply side, although one must not discount the difficulties which arise from the need for continued investment. Inadequate investment is linked to the problem of non-payment by customers, and may also relate to continued government controls over prices. As is the case in oil, foreign participation in the gas industry has thus far been marginal, the off-shore gas and oil projects on Sakhalin in the Far East being the major exception (Sagers, 1995b). This situation appears to stem from the way *Gazprom* has been able to wield its monopoly influence, persuading politicians that it is best placed to meet Russia's gas needs for the future. Since gas consumption now easily exceeds consumption of oil and coal in the country's energy balance, the continued success of this industry is of key significance to the economy.

In contrast to the gas industry, the coal industry has faced precipitate decline. In 1988, at the height of its output, the Russian coal industry produced 426 million tons of coal. By 1992, this had fallen to 337 million, and by 1996 to only 255 million. About three-quarters of output comes from Siberia, especially from the Kuznetsk Basin in West Siberia (producing almost one-third of Russian output in the early 1990s) and to a lesser extent from the Kansk-Achinsk lignite field in East Siberia. The problems of the coal industry long predate the post-Soviet era, suffering from rising costs as the more conveniently located fields began to be worked out, and denied the investment needed for revival. In the more recent period, the industry's problems were exacerbated by the break-up of the USSR, shortages of equipment, falling demand and strikes as coalfield workers protested against falling real incomes, the failure to pay wages on time and plans for restructuring the industry. Although coal prices were effectively freed in 1993, the industry has been subsidized in the effort to save jobs and output. However, the government now faces pressure from the International Monetary Fund (IMF) to restructure the industry and close unprofitable pits. Whether such restructuring will lead to a partial revival of the industry, or whether continued decline is inevitable, remains to be seen.

A surprising feature of the recent development of Russia's energy industry has been the failure of electricity output to fall to the same extent as output in industry overall. The implication of this is that electricity is now being used less efficiently. Factories may of course be underreporting their true output, as noted above, but it also seems likely that many factories are constrained to continue using electricity for heating, lighting and basic industrial purposes simply in order to continue to exist, despite the fall-off in production. Looking ahead to the next century, official policy, after long hesitating in the wake of the Chornobyl' disaster, now favours a further expansion in nuclear power. But it remains to be seen whether there is the available capital and the political will to carry this through.

Despite Russia's many energy problems, it is still in a far better position to resolve its difficulties than many of its neighbours who lack energy resources. Later chapters will describe how Russia has been able to make use of its energy advantage to advance its national interests beyond its frontiers.

The defence industry

It has already been noted that the military–industrial complex played a major role in the Soviet economy. Indeed, it is hard to exaggerate its significance. While the statistics are not entirely unambiguous, one estimate is that in 1988 the Soviet defence industry employed some 7.5 million people, a figure which rises to 12 million if the many ancillary activities are included (Cooper, 1991, pp. 12–13). The industry thus employed about 9 per cent of the total Soviet workforce, while the figure of 7.5 million would represent over 20 per cent of the industrial workforce. Since Russia had more than its fair share of the Soviet defence industry, its significance for employment there is commensurately great. According to another source, in the early 1990s (by which time, because of various administrative reorganizations undertaken under Gorbachev, the defence sector was bigger than it had been in the 1980s), the Russian defence industry employed between 6 and 7 million people, of whom perhaps 4.5 to 5 million worked in the machine building and engineering complex. The latter figure represents about one-half of total employment in the Russian engineering industry, or one-quarter of those employed in manufacturing industry (Noren, 1994).

It is very difficult from these figures to compare the relative significance of Russia's defence industry with those of other countries, if only because the output of the Russian industry was directed not only at the military but also at the needs of civilian industries and the consumer. Even so, it is clear that the scale of the problem of restructuring this industry in the post-Cold War era far exceeds that being experienced in the West. What makes matters even more difficult is the regional concentration of the industry. Thus, of the presently existing 89 administrative subdivisions of the Russian Federation, just seven employed 35 per cent of Russia's defence-related industrial workforce in 1990. These seven were: the city of Yekaterinburg and the surrounding Sverdlovsk

Oblast in the Urals, St Petersburg, Moscow city, Nizhniy Novgorod Oblast (Volga region), Perm' Oblast (Urals region), Samara Oblast (Volga region) and Moscow Oblast. Of these, the first three regions employed over 300,000 workers each, and the other four between 200,000 and 300,000. Eleven other regions employed between 100,000 and 200,000 workers each (Bass and Dienes, 1993). In absolute terms, then, such regions face the greatest problems of restructuring, but there are numerous regions, districts and cities where the relative importance of the defence industry is particularly great. Among these, the ten secret settlements involved in the fabrication of nuclear weapons (access to which is strictly controlled) are perhaps the most notable. But there are also many centres which are concerned with a whole range of defence-related activities that fall into this category. Where such centres are in remote locations, they are likely to face particularly severe problems of adjustment.

There is no doubt that a marketizing Russia can no longer afford a defence industry as big as that which existed in the Soviet period; in fact, as noted in an earlier chapter, it was the very size of the military machine which helped to bring about the downfall of the USSR. Since 1991, there has been a sharp fall in government orders for defence equipment, and the contraction in the defence and engineering industries has been particularly marked. One attempted solution has been to try to expand export sales but the world market for arms is limited and extremely competitive. This cannot, then, be more than a token solution. Another possibility is to convert defence plants to civilian production. Since such plants have been responsible for a considerable amount of civilian production in the past, there is already a track record in this area. However, in practice this policy faces considerable obstacles (Noren, 1994; Sanches-Andres, 1995). Quite apart from the technical problems involved, experience in the rest of the world tends to show that the management and working practices of defence firms are not easily adaptable to the requirements of the market. In Russia, where conversion has enjoyed meagre success to date, the problems have included: a very rapid decline in defence orders; a sharp reduction in demand for civilian engineering output; a chaotic financial situation; and weak and erratic state support for conversion (Noren, 1994). A further problem is quality: Russian's engineering products seem to find difficulty competing on the world market, and may face increasing competition at home from foreign imports. In only a few sectors is quality less of a problem – there has been some foreign interest in the aerospace, aviation, computer and telecommunications industries, for example. Privatization got off to a slow start in the defence industry, and there has been some organizational change (such as the formation of financial–industrial groups incorporating both manufacturers and financial institutions). But there are real doubts about how far privatization has been genuinely market-oriented, and how far designed to protect existing patterns of production with government subsidy.

In summary, Russia's future success as an industrial power must depend to a considerable degree on the success of its engineering industry, and this in turn depends on tackling the problems of the defence sector. Clearly, it is going to take many years to solve these problems.

Services

The lack of producer and consumer services in the command economy, together with the inadequacy of consumer goods production, has been commented on already. Russia's new, marketized economy requires many services which were unnecessary, or deemed to be so, under communism, while consumers are now demanding goods and services which were previously unavailable. The scene is thus set for a big expansion in services.

The available statistics, and common observation, both suggest that this is already happening. However, the official statistical service is able to record only part of the activity, omitting some of it entirely and recording other aspects in such a way as to obscure their significance. In terms of the official data on GDP, for example, wholesale and retail trade accounted for 8.8 per cent of GDP in 1989 but 16.6 per cent in 1994. Analogous figures for 'other' activities (including public administration and public and marketed services) were 20.1 per cent and 27.6 per cent respectively. In employment, whereas that in 'other' activities in 1994 was 91.3 per cent of the level in 1990 (compared with only 78.5 per cent in industry), employment in wholesale and retail trade had actually grown by 3.4 per cent over the period (see table 4.5). However, these data certainly underestimate the real extent of the changes, since much 'service' activity goes on unofficially.

Another suggestive set of statistics are those on small businesses, most of them new and privately owned, and engaged mainly in the provision of services of various kinds or in consumer goods production. *Goskomstat* recorded 842,000 small businesses in Russia at the beginning of 1997. In 1995, the largest concentrations were to be found in Moscow, with 20 per cent of Russia's small businesses, and St Petersburg, with 7.8 per cent.

The advanced capitalist economies have witnessed a considerable growth in producer services (which include such activities as accounting, advertising, financial services, management consulting, insurance, real estate, legal services) over the recent past (Daniels and Lever, 1996), but they were almost completely absent in the command economy. They are now beginning to burgeon in Russia. However, because *Goskomstat* still abides by the old categories in recording economic activity, it is very difficult to trace their development precisely. Statistics assembled by Gritsai indicate recent relative rises both in the Russian Federation as a whole and in Moscow – in particular in such activities as wholesale and retail trade, computer services, real estate activities, 'commercial activity promoting the market' (which embraces numerous producer services), finance and insurance, and administration and government (which also embrace some producer services) (Gritsai, 1997). Gritsai reports a 129 per cent growth in employment in Moscow's financial sector in the three years to 1993, and 30 per cent in wholesale and retail trade. The outstanding characteristic of the geography of producer services at the present time is their overwhelming concentration in the capital. Thus, according to Gritsai, producer services accounted for between 6.7 and 10.7 per cent (depending on one's definition of producer services) of the capital's total employment in 1993, much higher than in any other city, and between 31 and 45 per cent of the nation's employment in this sector was concentrated in Moscow. Moscow's role in the new banking sector is predominant, with about 35 per cent of all banks, 40 per cent of bank branches and over 80 per cent of all bank assets concentrated there. The focus on Moscow is hardly surprising, given the city's enormous advantages in international access, proximity to government and the fact that much of the international business activity is located there.

As the Russian economy develops, and regional economies begin to assert themselves, the presently dominant position of Moscow can be expected to be modified. Already a number of regional financial and service centres are beginning to appear, often with international linkages. These include St Petersburg in the Northwest, Krasnodar in the North Caucasus, Yekaterinburg in the Urals, Novosibirsk in West Siberia and Vladivostok in the Far East. Other potential centres include Nizhniy Novgorod, Samara and Kazan' in the Volga region, and some ports and border crossings.

Transportation

A modern industrial economy will inevitably require a modern and efficient transport system, and this will be particularly important in a country of the size of Russia. Unfortunately, as in so many other areas, the system inherited from the Soviets is neither modern nor efficient. The Soviet transportation system was designed to serve the needs of the command economy, and it reflected many of the deficiencies of that economy. It had a number of peculiarities. First, transport was developed to serve the USSR as a whole rather than its constituent republics and regions, and most of it was managed on a modal basis without regard to republican boundaries. Its principal purpose was to tie the disparate parts of the all-Union economy together, linking often highly specialized regional economies into one whole. One of its most characteristic features was the massive movement of raw materials and finished products over very large distances. This reflected the priorities of the command economy (emphasizing heavy industry), the failure to conserve resources with economic growth, the unbalanced nature of regional development and the tendency towards economic autarky. The average length of haul of goods in the USSR was probably the longest in the world, reaching 643 kilometres on all modes of transport by 1990, and 963 on the railways. Increasing demand led to enormous pressure being placed on most transport modes and to frequent bottlenecks.

A second peculiarity was the fact that, despite the much vaunted claim that the USSR possessed a 'unified transport system', the different transport modes rarely cooperated with one another and even found it within their interests to compete. Like other parts of the Soviet economy, transport development was uneven and to some degree even chaotic, reflecting the sectoral interests of the transport agencies rather than the system as a whole. There were limited incentives to seek efficiencies, to cut costs (not only were there many subsidies, but the Soviet pricing system made the calculation of real cost difficult if not impossible in many cases) or to take a long-term perspective. Indeed, the tendency, even after Stalin, was to regard transportation as an unfortunate necessity to be pared to the bone wherever possible. The economic planning system worked in such a way as to discourage investment in new capacity or equipment and to make ever more intensive use of the existing infrastructure and capital stock. Transport agencies also had little interest in seeking new business or in innovation – as with other parts of the command economy, their task was to fulfil their plans, usually expressed in crude, quantitative indicators. Many parts of the system were run near to the limit of capacity: hence the pressures noted above. Unfortunately, towards the end of the Soviet era, such pressures were enhanced by the shortage of capital and its diversion to projects like the Baykal–Amur Railway and the building of pipelines for northern energy development. The failure to invest sufficiently helped to precipitate a crisis on the railways in the late 1980s and fostered the more general transport problems which plagued the Soviet economy in its last years (Kontorovich, 1992).

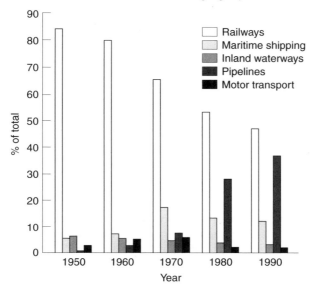

Figure 5.9   Russia: freight transport by mode, 1950–90.

A further peculiarity of the Soviet transport system, and one related to low priority, was continued reliance on the railways for freight transport (figure 5.9). In the Stalin period it made sense to rely on the railway network inherited from the tsars, improving it where necessary, rather than augmenting it with other modes. This attitude continued thereafter, although the government also began to invest in pipelines because of their advantages for transporting oil and gas, and in road transport for short-distance movements. Reliance on the railways, however, may not give the flexibility which a modern economy requires, and many commentators now advocate a switch towards roads.

A final oddity worthy of mention is the effects of the autarkic policies of the Stalin years, which helped to produce a continental, inward-looking transportation system rather than one oriented towards coasts and international frontiers (as the tsarist one had been to some extent). The autarkic policy was modified from the 1960s (the most obvious example on the railways being the Baykal–Amur Railway, connecting parts of East Siberia and the Far East with the Pacific – see figure 4.2). But it still meant that the former USSR was less well equipped with the infrastructure needed for international trade than might have been expected.

Table 5.2   Freight turnover in Russia, 1990–6 (billion ton-km)

	1990	1991	1992	1993	1994	1995	1996
Total	**5,890.6**	**5,457.4**	**4,697.8**	**4,157.6**	**3,566.5**	**3,532.6**	**3,358.1**
Rail	2,523	2,326	1,967	1,608	1,195	1,214	1,129
Road	68	65	42	53	36	31	25
Pipeline	2,575	2,404	2,146	2,019	1,936	1,899	1,913
Sea	508	464	405	373	311	297	220
Waterway	214	196	136	103	87	90	69
Air	2.6	2.4	1.8	1.6	1.5	1.6	2.1

Source: Rossiyskiy statisticheskiy yezhegodnik (1996, p. 586), Russian Economic Trends (1997, no. 1, p. 106).

In summary, then, the transportation system inherited from the past was built to serve an economy very different from that now developing under market conditions, with an infrastructure and equipment ill-adapted to the new situation, in dire need of new investment and designed to bind together the entire territory of the USSR, a country which no longer exists. It will take years to mould the system to meet the needs of modern Russia. All we can do in the remainder of this section is to mark the changes which have been occurring since the collapse of the command economy and to note some pointers for the future.

The transportation system could hardly escape the economic troubles of the later Gorbachev years and of the Yeltsin years in Russia. Indeed, as suggested already, transport was a significant contributor to those economic difficulties. Already suffering from inadequate investment, the system's performance was hampered by the negative consequences of Gorbachev's economic changes, partly because of the direct impact of the reforms on transport itself (resulting in a good deal of confusion) and partly because of their dire effects on the rest of the economy. In addition, towards the end of the Gorbachev period, transport began to suffer from the consequences of national breakdown: strikes, blockades due to ethnic strife and autonomist movements, rising crime and other difficulties. The result was a rapid deterioration in performance, which has continued into the Yeltsin era (table 5.2). Ironically enough, the rapidly declining demand for transport services proved a mixed blessing; if this had not happened, the system might have broken down entirely. The whole situation was exacerbated by the break-up of the USSR in 1991, as republics competed to claim as much of the transportation infrastructure and related property as they could and strove to reduce their dependence on one another. Part of the rationale for the establishment of the Commonwealth of Independent States was to overcome the disruptive effects on transportation and other vital inter-republican linkages of the dismemberment of the Soviet state. Unfortunately, mutual suspicion has reduced the degree of cooperation which might have seemed desirable on economic grounds, and transport has suffered accordingly.

Since 1991, Russian transport has been operating within the context of a developing market system (North, 1995, 1997). Transport agencies no longer simply respond to state orders as in the past, but increasingly respond to market signals, and they are now expected to pay their way. Thus, with the decline of traditional heavy manufacturing industry, transport is increasingly moving away from its traditional role of moving vast quantities of raw materials huge distances to feed such industries and towards other roles: the movement of raw materials and sellable manufactures for export, servicing the newer and more flexible activities which have accompanied the rise of the market, trade, tourism and other activities. All this requires a great deal of readjustment, which is going to be very difficult given the shortage of capital. As we have seen, not only is much of the infrastructure in need of renewal, but a great deal of it falls far short of what a modern economy requires. The lack of an adequate road network is a major problem for rural development and sometimes for industry and international trade; ports and airports lack modern facilities; outmoded and substandard ex-Soviet equipment is often costly to run and may not be versatile enough to put to new uses. Tariffs are now having to rise to raise the capital needed for investment, but this can have a detrimental effect on economic recovery. Thus far foreign capital investment has been restricted to nodes and activities having an international dimension.

While the attempt to seek profits will make transport agencies more efficient in a narrow sense, the effects may not always be in the national interest in the long run. This is because transport always has a social and political as well as an economic role. Thus there have already been many cutbacks in services across the country, which will inevitably exacerbate the development problems of the affected regions and exaggerate the fact of regional inequality. Removal of many government subsidies has had a dire effect on large parts of the north and east and threatens to undermine their economies completely. To some extent, the reorientation to the international market has been at the expense of domestic development, or at least threatens to become so in future.

As with other areas of the economy, the Russian government is obviously hoping that a reorganization of the transport system, principally by means of privatization, will make the system more efficient, flexible and aligned to the needs of a rapidly changing economy. Certainly there has long been a case for moving away from the inflexible Soviet approach to administering transport by mode, allowing for the establishment of intermodal companies and other flexible arrangements. Thus far, Russian railways remain in government hands and are controlled by the Ministry of Railways in Soviet fashion, though they are required to operate in a market environment. Other modes are overseen by the Ministry of Transport, but there has been considerable privatization except where the national interest seems not to favour it. As in industry, privatization has often been undertaken for defensive rather than entrepreneurial reasons. There has also been a rise of new private companies in aviation, road transport and elsewhere. Reorganization is an ongoing process, and the problem of reconciling private and public interests remains.

One area where the government is destined to have a permanent role is in transport relations with other countries. The new international borders with the other post-Soviet states have already resulted in the establishment of customs posts and ancillary facilities and in a reduction in the number of official crossing points. This necessarily has a hampering effect on trade. There are many issues of cross-border cooperation to be ironed out, including transit rights. Because of problems with some of its neighbours, Russia is now favouring the development of its own ports and routes to ensure access to the world's oceans and communication lines, even though this does add to overall costs. Thus, by building a new port at Luzhskoye Bay near the Estonian border, expanding facilities at St Petersburg and Vyborg, and diverting some trade to Murmansk and other northern ports, Russia is hoping to escape its previous dependence on the ports of the Baltic states. Similarly, port expansion on the Black Sea and the Sea of Azov may help to avoid dependence on Ukraine. Some of Russia's neighbours are likewise planning their own alternative routes, and thus the former Soviet transportation system is gradually becoming an historical phenomenon.

It has been seen that what Russia has inherited from its Soviet predecessor in the way of transport provides only a fragile basis for its transition to a market economy. What is needed is a much more flexible, modernized and complex system, which will somehow bring the benefits of emerging capitalism to the country as a whole. Whether this means such changes as a switch to road transport and away from the railways remains to be seen. Whether the capital for the momentous changes required will somehow be forthcoming also remains in doubt. And, finally, whether the economic role of transport can be reconciled with its undoubted political and social functions is also uncertain and is likely to prove a major challenge for the future.

Six

Saving the Environment

Any important economic change is bound to have environmental repercussions. This is as true of the economic transition which is now under way in Russia as it was of earlier phases of rapid economic change under the tsars and under the communists. The present phase is particularly difficult, however, not only because the country is facing new environmental challenges but also because it has to deal with the dismal environmental record of the past. How far Russia is successful in meeting these challenges is of interest both to its own population and to the world as a whole, since the changing environment of a country of Russia's size must have implications for the entire globe. The questions we shall be addressing in this chapter include: what environmental problems has Russia inherited from its communist past and how successful are its policies to solve such problems; how likely is the Russian economic transition to be hampered by environmental difficulties of various kinds; what major environmental problems are afflicting which regions; and can Russia find some 'environmentally sustainable' form of development?

The Geographical Context of Soviet Environmental Disruption

Environmental disruption occurs as a result of the impact of human activity upon nature. A major goal of both the later tsarist governments and the Soviet government was industrialization, and this was bound to produce environmental problems of various kinds. However, the particular kinds of problems which began to arise during the Soviet period, and which became more and more apparent as time went by, bore a close relationship to the command economy and its operation as described in earlier chapters. The decision to emphasize the development of heavy industries was bound to make significant demands on resources of energy, minerals and other kinds of materials. Such industries have almost always been of the polluting kind. The desire to industrialize as quickly as possible was likely to reduce the possibility of taking often subtle environmental considerations into account when considering new projects or expanding old ones. And the tendency to construct gigantic plants and installations in order to gain economies of scale often ensured that environmental risks were maximized over quite small areas. Of course, it was in the earlier stages of industrialization under Stalin that such policies reached their apogee, but similar tendencies were apparent throughout the Soviet period.

It is perhaps not entirely surprising that, in a country the size of the USSR, there was a tendency to think of resources as superabundant and perhaps even inexhaustible. The inclination to think of Siberia as a storehouse of resources, for example, long predates the Soviet era (Diment and Slezkine, 1993). In such a situation there seemed little need to worry about conservation and cutting down on waste. Although attitudes were modified with time, the inclination to classify resources as either 'exhaustible' (some of which, like vegetation and wildlife, were nevertheless seen as 'renewable') or 'inexhaustible' (mainly air and water resources) long characterized the Soviet view.

Marxism-Leninism taught that the laws governing nature are funda-mentally different from those governing human society (hence physical and human – or, as the Soviets called it, economic – geography were two different subjects). There could be no question of human development being somehow seriously constrained or retarded by the physical envi-ronment; thus there was little sympathy with any concept of environ-mental determinism (Hooson, 1959). In both political and academic parlance there was a good deal of emphasis on 'transforming nature' and moulding it to suit human needs. The feeling that in Russia nature, left to itself, might engender particularly severe obstacles to human development long predated the 1917 Revolution. This may have encour-aged a belief in possibilism on the grounds that its opposite would condemn the country to permanent backwardness (see Mazurkiewicz, 1992, pp. 26–32; Bassin, 1993).

Under Stalin the belief in the pliability of nature reached almost ludicrous extremes, with extravagant professions of faith in the potentials of science and technology, and large-scale plans being promoted to transform aspects of nature and thus to rectify its many 'defects' (Burke, 1956). Few such plans came to fruition. In fact, it is probably inaccurate to accuse Stalin of 'scientism' (i.e. an exaggerated belief in the potentials of science and technology). In the USSR science was subordinated to the political priorities of the regime. This came clearly to the fore during debates leading up to the first Five Year Plan, when the advocates of strict, scientific planning were routed by those who believed that sheer will and determination would conquer all – what Naum Jasny called 'Bacchanalian planning' (quoted in Hosking, 1992, p. 151). Some years later, N. I. Vavilov's careful genetically based research into the breeding of high-yield, drought-resistant crops was condemned in favour of the much more optimistic and wildly fanciful theories of Trofim Lysenko (Harris, 1988). Towards the end of Stalin's period, the extensive system of nature reserves, which existed for purposes of preservation and scientific research, was drastically reduced for purely utilitarian reasons (Weiner, 1988). In Stalin's 'war on nature' science might prove useful, but even science could not be allowed to stand in the way of the party's will.

Development plans under both Stalin and his successors potentially affected the entire territory of the USSR. The demands of industrialization set in motion an endless search for new resources, a search which was also motivated by the policy of economic autarky. The USSR's great rivers were also to be tamed and exploited for navigation, hydro-electric power, water supplies for industry and urban populations, and fisheries. The fact that the former USSR's territory is unevenly endowed with water resources fomented an obsession with the possibilities of inter-basin water transfers, which might also lead to climatic modification. Such was the potential expense of such projects that very few were actually undertaken, but the environmental implications of some of the more ambitious schemes were awesome (Micklin, 1986). As it was, the withdrawal of river water for irrigation, industrial and other purposes, together with related hydrological schemes, did help to produce some environmental disasters, of which the Aral Sea crisis is perhaps the best known (Micklin, 1992). A further territorial problem, or perceived problem, was the fact that arable farming and most human settlement were confined to about 10 per cent of Soviet territory, with perhaps a further 15 per cent usable for pastoral farming. Given the problems of Soviet agriculture, which became especially severe after collectivization, there were continual attempts to expand this 10 per cent: drainage schemes and similar land improvement projects in the north, irrigation and dry farming in the south. Unfortunately, such projects were not always undertaken with due regard to their environmental consequences, some-times leading to soil erosion, salinization or other phenomena detrimental to agriculture (Feshbach and Friendly, 1992).

It was in fact at the margins of human settlement and on the peripheries of the Soviet realm that policies of environmental improvement and economic development had some of their most deleterious consequences. These are regions where the natural environment is generally fragile and where the human impact can have particularly far reaching and severe consequences. Across the north, for example, Soviet policies of resource development have had serious effects on air quality, water purity and the productivity of land. In northern environments air pollution can travel vast distances, producing the phenomenon known as 'Arctic haze', seriously reducing sunlight and destroying vegetational complexes which may already be existing on the edge of viability in an extreme environment. The industries of the Urals, northern processing plants at places like Noril'sk and Igarka and other northern developments are widely regarded as responsible for much of the atmospheric pollution which has been detected in the Arctic. According to one source, the Noril'sk metallurgical combine in 1991 released almost 2.4 million tons of pollutants into the atmosphere. The sulphur oxide content of the air in the city is forty times the officially permissible limit (Scherbakova and Monroe, 1995, p. 69). Pollution from Noril'sk has had an extremely detrimental effect on the reindeer pastures of the Taymyr Peninsula and other vegetational complexes. What makes atmospheric pollution additionally hazardous across many parts of Siberia and the Far East is the winter high pressure system, which traps stagnant bodies of air in valleys and depressions beneath temperature inversions. Smog associated with exhaust emissions, ash, soot and carbon monoxide pollution from thermal electric plants results from chemical reactions in the atmosphere and can have devastating effects on human health. The pollution of rivers, lakes and reservoirs is another problem of northern development. Again, this can have serious and long-lasting effects because of fluctuations in river flow, especially in parts of eastern Siberia, and because of the reduced activity of aquatic micro-organisms in the bleak northern environment which can normally be relied on to break down organic pollutants. Water pollution can thus add greatly to the difficulties of securing supplies of fresh water for human consumption. Other much-cited problems of northern development include long-lasting damage to the permafrost, destruction of reindeer pasture and other kinds of vegetation from natural gas, oil and pipeline development (notably in northwestern Siberia) and the depletion of forests, which are slow to regenerate in harsh environments, through a combination of clear-cut logging methods, fires and atmospheric pollution.

Some of the most devastating environmental consequences of Soviet development policies in the semi-arid and arid regions of the south have been felt in Kazkhstan and Central Asia, outside the territory of the present-day Russian Federation. Soviet determination to boost cultivation of cotton and of certain other warmth-dependent crops led to continued efforts to expand the irrigation system and to ignore the consequences for the environment. The much-publicized results have included water shortages, secondary soil salinization in many places and the disastrous shrinking of the Aral Sea, with dire repercussions on neighbouring territories. Within Russia itself, however, unwise development in semi-arid regions has also produced undesirable environmental changes. These include widespread pollution of lakes, reservoirs and marine environments (notably, the Sea of Azov) and serious water shortages. Also significant has been soil erosion and deflation on the fragile but extremely fertile black earth and associated soils. This has a long history in the region but was intensified through the improper cultivation practices and general intensification attempted in the Soviet period. Perhaps the most well known instance is in the Virgin Lands region, the area of northern Kazakhstan and parts of West Siberia which, being mainly semi-arid, were cultivated for the first time in Khrushchev's day to boost output of food. The pressure to produce immediate results often allowed for little consideration of environmental niceties, and many areas were subsequently ruined by wind and water erosion.

The Political and Economic Context of Soviet Environmental Disruption

Before we go on to consider some of the more specific environmental problems inherited by the Russian Federation from the Soviet period, it is worthwhile saying a few words about the particular political and economic forces which helped to bring about those problems. This will enable us, later in the chapter, to compare those constraints with some rather different ones which are beginning to arise today.

Something has already been said about ideology, and there is no need to dwell much further on that here, particularly since the influence of ideology on actual policy-making in the Soviet period was by no means as straightforward as is sometimes imagined. Two additional points can be made, however. First, it is probably true to say that Marxism-Leninism, with its materialistic underpinnings, tended to encourage Soviet leaders towards a pragmatic attitude to the environment and away from a Romantic view. In other words, Soviet leaders tended to think technocentrically rather than ecocentrically – the environment was to be preserved or developed for the sake of human society and not for its own sake. Of course, this did not really provide ideological justification for Stalin's environmental extremism – Lenin was notably more cautious in his attitude (Schmidt, 1971; Kuznetsov and L'vovich, 1971, p. 32; Pryde, 1972, pp. 13–15). But it did tend to mean that Soviet leaders had few doubts about the basic rightness of their policies of economic development, despite the inevitable environmental consequences. They certainly would not have shared the sentiments of some ecologically minded commentators, who doubt the environment's stability in the face of the assaults of industrial civilization. Environmental problems, the Soviets tended to assert, were essentially the product of capitalism. Where they occurred under socialism, it was because of the incorrect application of basically sound principles, not because socialism itself was heading in the wrong direction.

A further ideological point is that Marxism-Leninism discouraged the idea that economic value can be produced by something other than by the application of human labour. Thus minerals in the ground, the land, air and water have no inherent value in economic terms and, since they are a 'free gift of nature', ought not to belong to anyone or yield rent. This made the Soviets reluctant to charge for the use of land, water and so on, arguably reducing incentives to use them sparingly.

Certainly more important than ideology in providing a context for Soviet environmental disruption was the character of the political system. As we have seen in an earlier chapter, the Communist Party was a highly autocratic body which placed a high premium on economic success. Environmental considerations were not to be allowed to hinder that success or to slow down growth rates. Politics influenced the party's attitude towards environmental transformation. Huge, prestige projects, having maximum environmental impact, not only would save money because of their scale but were a very visible symbol of economic achievement.

Ever since the civil war, there had been a close relationship and interdependence between the party and the military machine. Military might required a buoyant economy and the environment had to play second fiddle to the needs of the military. This meant not only enormous investment in a nuclear capability and other sophisticated weaponry, with all the attendant environmental hazards, but also a supporting civilian nuclear power programme. The risks involved in such developments were ignored both by officialdom, who no doubt believed that the risks were small and containable, and by a public kept largely in the dark. Joining the vested interests which supported the Soviet regime as time went on were the major industrial ministries, largely the handiwork of Stalin. The ministerial system, with its supporting network of enterprises and institutions, also stood full square behind the drive for industrialization. The ministries enjoyed enormous power at both national and local levels, and little could contain their desire for growth or their appetite for resources.

The other side of the coin of the Soviet political system was the weakness of the forces arrayed on the side of the environment. The

Soviet system had an elaborate body of environmental law and a multitude of agencies researching into, reporting on and charged with the duty of protecting the environment. Environmental tasks and duties were also written into ministerial and enterprise plans and their fulfilment was obligatory. In practice, however, all this had limited effect. The law and the judiciary were unable to constrain the vested interests of the political system, enforcement agencies were weak and hard-pressed ministries and enterprises had little incentive to worry about the environmental sections of their plans. As Gustafson writes with regard to the attitude of ministries and their enterprises, conservation policy faced the three major obstacles of industrial inertia, high cost and an uncertain pay-off (Gustafson, 1989, p. 228). Fines, when levied for a breach of environmental regulations, were invariably low and rarely paid personally by those responsible. Also very weak were the regional and local authorities (soviets) who might normally have been expected to know their own areas and to contest policies leading to environmental disruption. Not only was the latter very difficult under Soviet conditions, but local leaders were often more concerned about gaining economic advantages for their fiefdoms, and thus currying favour with the ministries, than about trying to restrain their activities.

A major difference between the Soviet system and Western-type political systems was the difficulty in the former of developing environmental pressure groups. It is arguably these groups which have been primarily responsible for changing environmental attitudes in the West. In the Soviet Union their formation was inhibited by laws restricting freedom of association and by official control of the press and other media. Although frequently drawing attention to relatively minor breaches of environmental protocol, the press was not free to engage in fundamental criticism of official policy. The public was therefore largely ignorant of, or indifferent to, major environmental difficulties. An important exception was the celebrated case of Lake Baykal in the 1960s, when an *ad hoc* group of academics, intellectuals and others protested against the lake's desecration by industrial interests (Gustafson, 1981). The group's success was only partial, however. Later, environmental concern grew partly through the unofficial activities of dissidents. Eventually, under Gorbachev, and especially after the Chornobyl' accident in April 1986, environmental activism came out into the open as a means whereby republics, regions and ethnic minorities could attack a now weakened centre.

The final point to be mentioned in this section is the working of the economic mechanism under the command economy. Despite attempted economic reform, the command economy worked in such a way as to minimize the effects of economic 'levers' (costs, fines, profits) and to maximize those of administrative ones (especially plan fulfilment, mainly in terms of quantities of output). Inputs like labour, land and other resources were rarely priced effectively, if at all, and there was every incentive, in an economy of shortages, to take more than was needed and to hoard the surplus in case of future need. Resources were therefore used inefficiently, and there was an inbuilt tendency for the economy to grow 'extensively', by using ever more resources to feed commensurate increases in output, rather than 'intensively', increasing output by using a given quantity of resources more effectively. The latter would require innovative behaviour for which there was little incentive in a situation that paid dividends to those who fulfilled their plans and minimized risk. Although not, strictly speaking, a result of the economic mechanism, overuse of resources was also encouraged by the bureaucratic behaviour of ministries which preferred to rely on their own systems and facilities rather than on those of other ministries. This led to duplication or multiplication of effort on many occasions.

The Russian Environment in the Early 1990s

Mikhail Gorbachev's espousal of a policy of *glasnost'* ('openness'), to- gether with the experience of the disaster at Chornobyl', led to a consid- erable increase in the availability of environmental information towards the end of the Soviet era. A new agency, the USSR State Committee for Environmental Protection (*Goskompriroda*), was established in 1987, and soon this and other organizations began to issue new data. Using this information as a basis, it is possible to say something of the en- vironmental problems inherited by the Russian Federation in 1991. This cannot, however, be a comprehensive survey. For this readers are referred to the available sources (see Panel on the State of the Soviet Environment, 1990; Pryde, 1991; Feshbach and Friendly, 1992; Peterson, 1993).

The Russian Federation inherited a particularly dismal environmental record from its Soviet predecessor with respect to air pollution (Panel on the State of the Soviet Environment, 1990, pp. 403–9). Statistics issued by the former Soviet *Goskompriroda* suggest that total atmospheric emis- sions in the USSR in 1988 amounted to 78 or 79 per cent of those of the USA. Allowing for the fact that Soviet GNP in that year was only about 54 per cent of that of the USA, according to maximum Western esti- mates, this gives a rate of air pollution 2.5 times that of the USA per unit of GNP. Even more significant is the fact that Soviet industrial air emissions were 90 per cent of those of the USA, despite the difference in GNP. In some areas of the former USSR the situation was particularly worrying. Thus 1989 statistics suggested that 20 per cent of the Soviet urban population were living in 103 cities in which the level of at least one major pollutant exceeded the maximum permissible concentration by ten times or more on at least one occasion in the previous three years. Most of these cities are in the Russian Federation. A 1992 document lists 84 Russian cities where at least three pollutants exceed permissible limits by ten times or more (Pryde, 1995, p. 31). In fact, only 15 per cent of Russian urban residents are said to live in areas having levels of atmospheric pollution which fall within permissible limits (Kochurov, 1995, p. 50). The worst affected cities for individual pollutants have commonly been medium-sized cities with a heavy industrial base, particularly metallurgy, oil refining and chemical production. Russian cities in such economic regions as the Urals, the North and East Siberia are among the worst hit. However, for aggregate indices of pollution the former USSR's biggest cities are very significant The former Soviet *Goskompriroda*'s report on the state of the Soviet environment in 1988 listed 68 of the most polluted cities in the USSR. The list contained over half of the 'millionaire' cities and 44 per cent of those with over 500,000 people – a total of 16 per cent of the Soviet population. Half of the 32 biggest cities in European Russia in 1989 were on the most polluted list, including Moscow and St Petersburg (Leningrad).

Russia is characterized by a very uneven distribution of water supply relative to population, and has worrying problems of water pollution. Thus, according to the Ministry of Environmental Protection's annual Report on the State of the Environment in 1994, whereas West Siberia is supplied with 44.7 cubic metres of water annually per head of the population, East Siberia 136 and the Far East 297 cubic metres, the Urals has only 6.6, the North Caucasus 4.3, the Central Region 3.9 and the Central Black Earth Region only 2.7 (table 6.1). For the country as a whole, annual demand for water from all natural sources in the mid-1990s equalled about 2 per cent of renewable sources, but demand varies considerably by region and in some areas, especially towards the European south, can reach up to 40 per cent of supply. Bearing in mind that a considerable proportion of the available supply is needed to maintain river flow and keep lakes and inland water bodies supplied, such regions face real water deficits unless demand can be restrained. Annual demand for fresh water for agricultural, industrial and domestic use reached 73.2 cubic kilometres in 1996, although this had fallen from 102.2 in 1985. Water conservation measures, industrial decline and reduced demand for irrigation water because of rainfall levels were all responsible. Of the 73.2 cubic kilometres, 19 per cent was used for irrigation and agricultural needs, 53 per cent for industry and the rest for domestic and other purposes. Russia is, of course, far better placed

Table 6.1 Annual water supply to Russia's economic regions (thousands of cubic metres per year)

Economic region	Water supply (thousands of cubic metres per year)	
	Per square kilometre	*Per capita*
Northern	349	90.6
Northwestern	455	11.6
Central	232	3.9
Central Black Earth	125	2.7
Volga–Vyatka	576.5	18.2
Volga	503	17.3
North Caucasus	195	4.3
Urals	156.6	6.6
West Siberia	241	44.7
East Siberia	273	136
Far East	290	297
Russian Federation		28.5

Source: *Zelenyy mir* (1995, no. 31, p. 4).

for water resources than some of its neighbours to the south. Central Asia, with its growing population and lavish consumption of water for irrigation, faces particular problems, especially now that it can no longer rely on the rest of the USSR for help.

Potential water shortages are made worse by the problem of pollution (Panel on the State of the Soviet Environment, 1990, pp. 409–15). Pollution tends to be especially notable towards the south: according to one official ranking, the worst economic regions in the USSR for water shortages and pollution were the Volga region and Donets–Dnepr (Ukraine), followed by the North Caucasus, the Urals and Central Asia. Of the 24,642 million cubic metres of polluted water discharged into Russia's rivers and water bodies in 1994, 45 per cent went into the Caspian drainage basin, including 39 per cent which went into the Volga and its tributaries. The Caspian Sea off Baku and the sea's northern portion are particularly polluted, as well as the rivers running into it. The Sea of Azov basin received about 13 per cent of Russia's polluted discharge, much of it by way of the Don and the Kuban' rivers, which are among Russia's most polluted rivers. The Sea of Azov and neighbouring parts of the Black Sea are noted for their pollution problems, which have had a detrimental impact on tourism and other activities. Water pollution afflicts many other parts of Russia as well, including such notable water bodies as Lakes Ladoga, Onega and Baykal. No fewer than 20 per cent of the drinking water samples taken in 1990–4 across Russia failed to meet basic safety standards. Major causes of water pollution include industrial effluent, the seepage of fertilizers and pesticides from cultivated land, urban impacts and other processes.

Although the Russian Federation has the largest land endowment of any country in the world, there is considerable concern about land degradation and losses of land to various destructive processes aided and abetted by human action. The cultivated area, for example, has declined by about 15 million hectares since the mid-1970s, a reduction of about 12 per cent, but no doubt much of this is due to the cessation of cultivation on marginal lands. Given the generally poor performance of Russian agriculture, losses of land resulting from natural processes or its reallocation to non-agricultural uses are a cause for concern, as are reductions in fertility. In the past, losses to reservoirs and to urbanization were considerable, but these seem to have abated somewhat in the 1980s. Soil erosion remains a serious problem – about a third of the cultivated area was affected by wind erosion in the 1980s and almost 20 per cent by water erosion. Recent data indicate a loss of soil humus at an average annual rate of 600 kg per hectare, and 17 Russian regions have reported desertification processes. About a third of the cultivated area was affected by acid rainfall. Land improvement schemes like irrigation and land drainage projects can also have negative environmental side-effects. Another problem is the dumping of toxic and urban wastes: the disposal of various kinds of industrial residuals (hazardous chemicals, toxic metals, asbestos etc.), contamination of the land by pesticides, herbicides and different types of fertilizer and the treatment of municipal waste have all caused difficulties, despite advances in recycling and more efficient modes of waste disposal. Losses of land to mining and associated activity totalled about 1.1 million hectares in Russia in 1995, despite progress with reclamation (see Bond and Piepenburg, 1990). Destruction of northern habitats through construction projects, and pollution by oil spillages and other accidents, are also serious.

A particularly grave pollution problem for the Russian authorities is contamination by radioactivity (Pryde and Bradley, 1994). This results from the Soviet development of a major nuclear capability for military purposes in the tense years of the Cold War and the large-scale civilian nuclear programme. According to Pryde and Bradley, the major sources for nuclear pollution include: the operation of nuclear reactors and nuclear fuel reprocessing plants; the disposal of nuclear wastes; uranium mining and processing; nuclear accidents; the testing of nuclear weapons; the use of nuclear explosions for 'peaceful' purposes; and the use of radioactive isotopes in industry, hospitals, research facilities and similar locations. The most spectacular release of nuclear radiation into the atmosphere occurred with the Chornobyl' nuclear accident in Ukraine in April 1986, which polluted as much as 50,000 square kilometres of Russian territory with radionucleides and had an even more serious impact on Belarus' and to a lesser extent on Ukraine. In Bryansk Oblast in western Russia, an area of 310 square kilometres was contaminated with radioactivity at a level of up to 40 curies per square kilometre in 1993. Another serious case of radioactive pollution occurred at Chelyabinsk-65, a nuclear reprocessing centre in the southern Urals (Monroe, 1992). Here at the Mayak facility radiation escaped or was released into the environment over a lengthy period. The major accident known as the 'Kyshtym incident', which occurred in September 1957, was only the most spectacular of a whole series of events and leakages contaminating the area. Up to 500,000 people are believed to have been exposed to harmful doses of radiation. Other Russian sites which are associated with serious radioactive pollution include reprocessing and waste storage facilities at Tomsk-7 and Krasnoyarsk-26, nuclear dumps on the island of Novaya Zemlya in the Arctic and the adjacent Barents and Kara seas, Lake Ladoga, parts of Primorsky Kray in the Far East and other locations.

Russia faces many other environmental problems, such as the need to conserve biotic resources (including forests), wildlife and inland and coastal fisheries. On the positive side, however, Russia had over 19 million hectares of land under various forms of official protection in 1989, a number which has since risen to 38 million. An overall ranking of the 19 Soviet economic regions in terms of their degree of environmental disruption in the 1980s is given in table 6.2. This compares regions with regard to their degree of atmospheric pollution, soil contamination and erosion, water shortages and pollution and forest damage. It shows that the aggregate picture in Russia is most serious in the Urals, followed by the Central Region, the Volga and the Central Black Earth Region (figure 6.1).

A recent and more sophisticated way of depicting the geography of environmental disruption in Russia is the identification of critical environmental zones (Pryde, 1994). These would be zones where the presence of one or a combination of several environmental problems produces a situation hazardous to human health. Zones can vary in their degree of intensity. One map of the former USSR compiled in 1991 designated a number of critical zones within the Russian Federation. They included:

1 The Chornobyl' fallout region.
2 The Kuznets basin, suffering from air and water pollution and land degradation as a result of mining.

Table 6.2 Rank ordering of 19 Soviet economic regions by degree of pollution and environmental degradation at the beginning of the 1980s

Region	1	2	3	4	5
Russia					
North	2	12	8	2	13
Northwest	10	13	9	7	15
Central	6	10	3	6	4
Volga–Vyatka	15	11	4	18	17
Central Black Earth	9	3	4	14	6
Volga	8	6	1	12	5
North Caucasus	12	6	2	14	9
Urals	1	9	2	3	2
West Siberia	7	14	10	5	16
East Siberia	3	15	11	1	15
Far East	16	15	12	8	18
Other					
Donets-Dnieper	5	4	1	4	1
Southwest	11	7	6	15	8
South	17	5	5	17	12
Baltic	18	8	6	9	12
Transcaucasus	15	2	3	10	3
Central Asia	14	2	2	11	7
Belarus'	13	5	5	13	11
Moldova	20	1	3	16	10

Key: 1, air pollution per unit of urbanized area; 2, soil contamination and erosion; 3, pollution of rivers and water shortages; 4, damage to forests; 5, aggregate human impact.
Source: Panel (1990, p. 408).

3 The Urals industrial region, which has extreme air and water pollution from its many industrial activities and was also the site of the Kyshtym incident and associated nuclear contamination problems.
4 The Kola Peninsula, suffering from air pollution as a result of non-ferrous mining and metallurgy.
5 The Kalmyk Republic, with soil erosion and soil chemical contamination.
6 The Black Sea, with water pollution problems.
7 The Sea of Azov, with similar problems.
8 The Volga River, again with water pollution and problems deriving from hydro-electric installations.
9 The Caspian Sea, whose problems also derive from the oil industry.

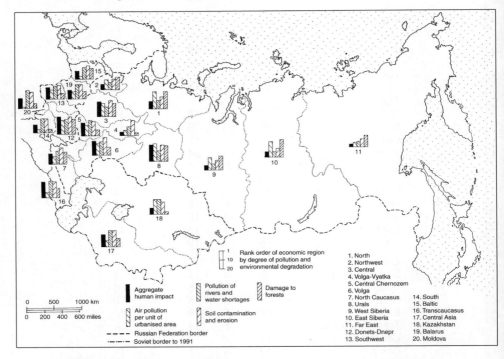

Figure 6.1 Major environmental problems in economic regions of the former USSR, late 1980s.

10 Lake Baykal, whose unique ecosystem has been damaged by timber processing and other activities.
11 Moscow city, with problems of air pollution (industrial and automotive) and radioactive waste.

The 1991 map also lists critical zones in other post-Soviet republics, including, of course, the Aral Sea. But certain problem regions in Russia, like Noril'sk and Novaya Zemlya, for some unknown reason do not appear as critical.

A second and more detailed map completed in 1993 identifies 60 environmentally 'very critical' locations and regions on the territory of the Russian Federation. Of these, 19 are in European Russia, 12 in the Urals, 6 in the North Caucasus, 6 in West Siberia and 17 in East Siberia and the Far East. The framers of Russia's 1991 environmental protection law claimed that 20 million people in Russia were living in 'zones with an extreme ecological situation' and 'ecological disaster zones' (Bond and Sagers, 1992, p. 470) (figure 6.2).

Environmental Management since 1991

Since 1991, the evolution of Russia's environment and attempts to manage that evolution have been taking place in a context which has been undergoing profound change. Rule by an autocratic Communist Party committed by ideology and political interest to industrialization practically at any cost has been replaced by a system open to democratic challenge and to a plurality of interests. Henceforth citizens who are dissatisfied with the course of environmental policy have resort to the ballot box as well as to other channels of democratic influence. From now on the activities of government, public and private organizations and individuals should be conducted within a framework of law. No longer should it be the case, as it was in the Soviet period, that powerful organizations or individuals, be they government institutions, private companies, military personnel or political parties, are above the law. Democratization also means that many more decisions should now be made at regional or local level, closer to the environmental issues which actually affect people's lives. People are now free, as they were not in the Soviet era, to speak out on environmental issues, form pressure groups and political parties to campaign on environmental problems and challenge vested interests. Education and the media are more likely to reflect environmental concerns than they were in the past. In other words, environmentalism is now provided with much more fertile soil in which to take root than it was even in the Gorbachev years.

Proponents of the market economy also point to environmentally positive aspects of the current transition. Under market conditions, they argue, most resources will have a price and firms will be forced to use them efficiently. Innovation will be encouraged in the quest for efficient resource use. Private ownership of assets is also regarded by many as a force promoting conservation and efficiency. Under the conditions now prevailing in the world market, moreover, the resource-demanding smokestack industries of the Soviet period are unlikely to survive and should give way to less polluting and environmentally unfriendly activities.

By no means all the post-1991 changes are likely to prove environmentally beneficial, however. Even elected Russian governments may behave in an authoritarian and environmentally insensitive way and the various checks and balances now in place may be insufficient to restrain

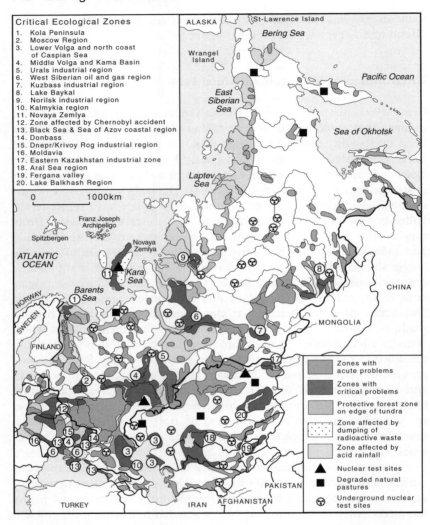

Critical Ecological Zones
1. Kola Peninsula
2. Moscow Region
3. Lower Volga and north coast of Caspian Sea
4. Middle Volga and Kama Basin
5. Urals industrial region
6. West Siberian oil and gas region
7. Kuzbass industrial region
8. Lake Baykal
9. Norilsk industrial region
10. Kalmykia region
11. Novaya Zemlya
12. Zone affected by Chernobyl accident
13. Black Sea & Sea of Azov coastal region
14. Donbass
15. Dnepr/Krivoy Rog industrial region
16. Moldavia
17. Eastern Kazakhstan industrial zone
18. Aral Sea region
19. Fergana valley
20. Lake Balkhash Region

Zones with acute problems
Zones with critical problems
Protective forest zone on edge of tundra
Zone affected by dumping of radioactive waste
Zone affected by acid rainfall
▲ Nuclear test sites
■ Degraded natural pastures
☢ Underground nuclear test sites

Figure 6.2   Acute and critical ecological zones at the end of the Soviet era.

them. The 1993 constitution gives great power to the executive, and it is by no means clear that the parliament or the courts will prevent infringements of the law. As in the past, economic success is still perceived as a cogent necessity by politicians and others. The environment may well be sacrificed to attain it. Unfortunately, Russian courts do not have a good record of independence, and they may well be overawed by powerful interests. The latter include government, private enterprise and also the military, whose influence may ultimately prove to be no less pervasive than it was in the Soviet era. Russia's nascent bureaucracy, often staffed by those who held office in the Soviet era, similarly leaves a great deal to be desired. Officials charged with the duty of upholding environmental standards nationally or locally may lack expertise, motivation or the necessary independence of mind to be effective (Peterson, 1995a). The devolution of power to regional authorities may not always mean more environmentally sensitive policies, since local politicians, no less than national ones, may well be happy to trade the environment for economic advantage. The public too may prove indifferent to environmental issues and prefer jobs to a safe and healthy environment for living. Indeed, there has been a sharp decline in environmental interest with rising economic difficulties since the fall of communism, and green policies and parties have made little headway. As one example of the change in attitude, nuclear power, which threatened to go into eclipse after Chornobyl', is once again on the agenda, and a big expansion programme was announced in 1993 (Marples, 1993). A final point is that the rise in criminality at all levels threatens to undermine legality and the well-being of the environment for the sake of private gain.

Since 1991, Russia has been equipping itself with the infrastructure necessary to combat environmental disruption. A major landmark was the Law on Protection of the Natural Environment, approved by President Yeltsin in December 1991. Although this is now to be redrafted to bring it into line with subsequent events, including the introduction of a new constitution in 1993, the law is meant to provide a framework for future environmental legislation. It is, in the words of two drafters of the law, Russia's 'ecological code' (see Bond and Sagers, 1992, pp. 463–4). The law establishes the basic principles which are to govern the conduct and management of ecological matters for the future. The major emphases, according to a report by Bond and Sagers, include: (a) defining the jurisdiction and responsibilities for environmental affairs at all levels, from national (federal) down to local; (b) specifying measures appropriate to different stages in the economic development process; (c) tailoring environmental protection strategies to individual economic sectors; and (d) strengthening the legal basis for public participation in decisions affecting environmental issues (Bond and Sagers, 1992, p. 464). In theory, therefore, the law should help to eradicate some of the jurisdictional confusion which was a characteristic of the Soviet period, increase the accountability of those involved in activities which disturb the environment and promote public consultation on environmental matters. Among the law's more interesting provisions is the requirement to establish procedures for implementing the 'polluter pays' principle, whereby licences must be acquired by those permitted to engage in environmentally disruptive activities and fines are levied on those who infringe the specified limits on those activities. Money derived from these procedures is to be paid into special ecological funds, which can then be used to clean up the environment. The law guarantees citizens the right to a healthy environment, and the right is made concrete by the recognition of sets of 'environmental quality norms' or standards (such as maximum permissible concentrations of pollutants, or MPCs) to underpin environmental protection. Other provisions include those which deal with the financing of environmental protection measures, the establishment of systems of ecological inspection, environmental monitoring and education, the resolution of environmental disputes, procedures for dealing with infringements of environmental law and specific provisions dealing with nuclear pollution and the activities of the military. A section concerning 'extreme ecological situations' (an unfortunate necessity in the light of Chornobyl' and other situations discussed above) lays down provisions for defining 'zones with extreme ecological situations' and 'ecological disaster zones', the latter being the more rigorous of the two. Another section deals with specially protected areas like nature reserves, national parks, resort regions and green belts. Finally, the section on international cooperation pledges Russia to cooperate with other states in securing a healthy global environment, a marked departure from attitudes prevailing during the Soviet era.

The 1991 law represents a laudable attempt to place environmental protection in Russia on a new footing after the dismal record of the Soviet years. But the success of the law greatly depends on how its provisions are translated into further legislation and how it is implemented in practice. Since 1991, a large body of legislation has been enacted. The most notable is the 1993 constitution, which accords environmental protection a significant role in the new Russia (*Konstitutsiya*, 1997, article 42). Further laws include those on mineral wealth (1992), wildlife (1995), the land code (provisional), protected areas (1995), principles of forest legislation, principles of health protection legislation and others. New laws to be enacted include a water code, a forest code and a law on federal natural resources. Other legislation, like the criminal code, and laws on business, taxation, local government and other matters, also contains environmental provisions.

Since 1991, when it replaced the former USSR State Committee for Environmental Protection as far as Russia was concerned, the principal organization responsible for conservation in Russia has been the Ministry of Environmental Protection and Natural Resources (now the State Committee for Environmental Protection, *Goskomekologiya*). This body exists at federal level and has regional committees at the level of the oblast and kray. The federal body is responsible among other things for long-range policy, setting environmental standards, managing plant and animal resources, overseeing nature reserves and administering international environmental agreements. It also issues an annual report on the state of the environment. The regional committees carry out instructions from the centre, collect environmental data, help to issue licences for resource use and enforce the law. At local level (cities, rayons), the state committee and the local authorities are jointly responsible for establishing bodies to oversee environmental matters. When the environmental ministry became solely answerable to the Russian government back in 1991, the idea was to subordinate the many other agencies with environmental responsibilities (like the State Hydrometeorological service, which collects data on air and water pollution, or the State Committee for Sanitary-Epidemiological Supervision, which monitors drinking water, food quality and other health-related matters) to it. However, this move was stoutly resisted by some of the organizations concerned, and hence there are grounds for believing that some of the overlap and confusion of responsibility which so plagued conservation policy in the Soviet period will continue.

Perhaps in partial continuation of Soviet practice, one of the duties of *Goskomekologiya* is to participate in the development of plans for environmental action by government. In 1994, for example, a presidential decree concerned the elaboration of a State Strategy for Environmental Protection and Sustainable Development. One of the provisions of this decree was to require the construction of a government Environmental Action Plan for the two-year period 1994–5. This was duly approved in May 1994. This was followed by a second Action Plan for 1996–7, linked with a proposed document known as the Concept of the Transition of Russia to the Model of Sustainable Development. The aim of the latter is to try to ensure that, under conditions of a market economy, Russia moves away from the environmentally disastrous course of the past.

It is already apparent that Russian environmental policy faces many problems which need to be overcome before the country can hope to eradicate the mistakes of the past and move on to an environmentally sustainable path for the future. One of the most pressing difficulties is that of money. Less than 1 per cent of the federal budget is currently spent on the environment, which is woefully inadequate for the tasks ahead. Neither will passing the major responsibility for solving environmental problems down to regional and local levels help much. Regional and local government also desperately needs money (some more than others) and frequently lacks the expertise which is available nationally.

Applications by regional and city governments for help from the newly established ecological funds are unlikely to be satisfied in times of financial stringency. Another problem, which can probably only be solved nationally, is the inadequacy of the available environmental information. An OECD report pointed out the lack of attention being paid to data quality, the absence of standardized procedures in data collection and handling, and endemic secrecy as major issues requiring attention (OECD, 1996). The report urges the creation of a unified state system of environmental monitoring. Major educational and media efforts are also required to change public attitudes towards the environment. When the problems of political infighting, bureaucratic muddle and judicial weakness are considered, it is possible to assert that Russia still has some way to go before it is likely to find a model of sustainable development.

The Russian Environment in the Transition

The current economic transition is having a marked effect on the state of the environment, ameliorating some kinds of environmental disruption and exacerbating others. Thus, in addition to the heritage of the past, the Russian government is having to cope with the environmental changes being produced by the new market economy. The present section considers these changes, giving some consideration to regional differences.

On the face of it, the situation with regard to air pollution seems promising. Stationary source air pollution for cities of the Russia Federation declined from 38.5 million tons in 1988 and 34.1 in 1990 to 19 million in 1997. These data mainly measure emissions from heavy industry, which has traditionally been the principal source of atmospheric pollution. The main reason for the reduction in the 1990s is deindustrialization in the wake of the economic reforms of Gorbachev and Yeltsin. The 41 per cent decline in harmful atmospheric emissions over the period 1990–6 compares with a 52 per cent decline in industrial output over the same period. It is, of course, very difficult to be confident about the data either on atmospheric pollution or on industrial production, for reasons which have been discussed above and in chapter 4. It seems likely, however, that pollution has declined less steeply than GNP. One probable reason for this is the need to keep certain industrial processes going even when the demand for the final output has fallen – technology, in other words, cannot respond perfectly to changes in the market. Another reason is the possibility that industrialists have been cutting expenditure on environmental protection in preference to other costs, like jobs or output. If this is true, it means that while atmospheric emissions have diminished, the intensity of pollution – the amount of pollution emitted per unit of output – has risen. Moreover, in 1993–4, emissions increased in the case of some 4,000 enterprises, or over 20 per cent of those for which data are available. Part of this increase results from growing export demand for the products of some polluting enterprises.

Much of the industrial pollution occurs within cities and, while data on individual cities are not easily obtainable, the available evidence suggests that atmospheric emissions have declined overall. Average concentrations of harmful substances in the atmosphere in 1994 exceeded maximum permissible limits in 208 cities with a total population of 64 million (43 per cent of the population). This compared with 231 the previous year. Eighty-three cities (with a combined population of over 40 million) had atmospheric pollution levels on individual occasions which exceeded maximum permissible limits by a factor of ten. This compared with 86 cities in 1993. In Moscow, harmful atmospheric emissions have fallen from 274 thousand tons in 1990 to 174 in 1995. A similar decline has been experienced in St Petersburg.

Offsetting these gains, however, have been sharp increases in mobile source air pollution in cities, especially from motor cars. Total emission of harmful substances into the atmosphere from road transport in Russia equalled about 13.5 million tons in 1994, but about 56 per cent of this came from trucks. Chapter 5 described the recent decline in freight transport, and this is reflected in atmospheric emissions: in 1993, road transport emissions equalled as much as 19 million tons. However, emissions from cars have been growing apace, especially in big cities – there was a 58 per cent increase in the number of private cars in Russia between 1990 and 1996. Typically, road transport contributes 40–50 per cent of atmospheric pollution in cities, but in 150 cities pollution from cars significantly exceeds that from industry. In Moscow, where road transport contributes up to 80 per cent of atmospheric pollution, there was a doubling of mobile source air pollution over the 1990–6 period.

If the situation in air pollution is mixed, that in water pollution seems distinctly less promising. Again, superficially the situation seems to be improving. Thus discharge of polluted water to surface water bodies in 1997 equalled 21.5 cubic kilometres, compared with about 27 in 1993 and for several years previously. About 33 per cent of this pollution comes from industry, which is the basic cause of the decline, and 61 per cent from the domestic economy. Agriculture's contribution to water pollution comes principally from run-off containing pesticides and other impurities, but agriculture's difficulties have resulted in less investment being available for chemical products. However, agricultural drainage is largely excluded from the pollution statistics, and the available evidence points to a progressive deterioration of water quality, with increasing numbers of surface water bodies and groundwater supplies being polluted. Ninety cities and settlements experienced deterioration in the quality of their drinking water supplies from groundwater in 1996. Among factors contributing to water pollution problems are poor maintenance of water treatment and sewage facilities, and inadequate policing of the regulations on waste water discharge and the management of catchment areas. About 50 per cent of Russia's population drinks contaminated water.

The situation with regard to land pollution and degradation is likewise a continuing source of concern. Only 10 per cent of the toxic substances produced by industry are processed and rendered harmless, and only 45 per cent are utilized. The rest are simply dumped. The agricultural area continues to contract because of land degradation by various means and assignment to other uses. Such losses are inadequately monitored at the present time. The area of land suffering from radioactive pollution remains fairly stable.

Russia has developed a number of state programmes to address some of the worst environmental consequences left by the years of communist development. It is too early to assess the effectiveness of such programmes, except to note that several of them have suffered through shortages of funds.

Building on the regionalization of economic change outlined in chapter 5, it is possible to say something about the kinds of environmental impacts which different regions of Russia are experiencing in the current economic transition (see also Peterson, 1995b). Like the regionalization scheme itself, the environmental impacts cited involve the broadest generalization and there will be many exceptional situations within each region.

Agricultural regions Russia's breadbasket, situated mainly in the south of European Russia, is suffering from water shortages, water pollution and land degradation. Inmigration in the wake of the break-up of the USSR and of the region's climatic advantages raises problems of adequate provision of drinking water, sewage treatment and disease control. Although these regions have tended to be somewhat conservative in their approach to economic reform, some parts may benefit from their climate or location near the sea coast or Russia's international frontier, and their economic revival could exacerbate their pollution problems. Recent reductions in state investment into soil protection have exacerbated erosion, desertification and similar problems, but reductions in expenditure on the use of agrochemicals have generally beneficial environmental effects.

Gateway and hub regions Moscow, St Petersburg and other centres have benefited from the environmental effects of deindustrialization (reductions in stationary source air pollution, less reliance on military production) and the trend to small-scale activities, including services. But some of the benefits seem likely to be cancelled out by problems associated with affluence (for some), such as pollution by motor cars and the generation of increasing amounts of solid and liquid waste, some of a much more sophisticated kind than was common before. Growing social inequality may also produce problems of its own (disease, higher death rates, crime and social disorder). Another growing problem is that of suburbanization, swallowing up valuable agricultural land or other open space, and requiring provision of running water, sewerage and other services. Expanding cities make ever greater demands on recreational resources, water supplies, space for waste disposal and other needs.

Resource regions Those parts of northern and eastern Russia which can sell their natural resources at home or on the world market face pressure to develop those resources in a cheap and not necessarily environmentally friendly way. The exploitation of energy resources, minerals and forest resources can have serious environmental side-effects, especially given a 'frontier mentality' which seeks for quick profits. Accidents like oil spillages, interference with natural processes in regions with fragile ecologies, poaching and other human impacts can have serious, long-term consequences. Some resource regions are characterized by the presence of resource processing industries (non-ferrous metallurgy in East Siberia, pulp and paper manufacture in the European north), whose deleterious effects on the environment look set to continue if such industries benefit from world market demand.

Old industrial regions The USSR's old industrial heartland, often referred to as the 'Rustbelt', has benefited environmentally from deindustrialization but must cope with some of the worst instances of the environmental heritage of the Soviet era. Where Soviet-era industry survives because of government subsidy or world market demand (as is the case with some metallurgical production in the Urals or chemical and petrochemical production in the Volga region), the environmental consequences will continue to be serious, unless ameliorative measures can be introduced.

'High-tech' regions The uncertain future of some industrial regions with a scientific and research tradition was noted in chapter 5. If unsuccessful, such regions will benefit environmentally from deindustrialization. If more successful, they are likely to continue to suffer the effects of industrial development or perhaps even some of the 'problems of affluence' described for gateway and hub regions.

The International Dimension

These days the condition of the environment is no longer the concern of individual states, as more and more environmental issues take on regional and even global dimensions. The USSR was often a reluctant participant in international environmental agreements, but Russia has largely inherited the agreements and treaties which were signed at that time. By 1996, Russia was a signatory to some 22 multinational agreements on environmental monitoring and improvement, and had entered into 33 bilateral agreements on the environment with states in western Europe, North America and Asia. As well as taking part in programmes of a general nature, Russia participated in a number of regional programmes, such as those on the Black Sea, the Baltic, the Barents Sea, the Arctic and Antarctica. A signature on international agreements, of course, does not always mean enthusiastic implementation. Nor has Russia always been forward in acknowledging the existence of global or regional issues – it has been slow, for example, to recognize the seriousness of such problems as global warming, acid rain and Arctic haze. Closer to home, Russia has been working with the other post-Soviet states on environmental issues within the framework of the Commonwealth of Independent States. An interstate council has been established for this purpose. As well as discussing issues of environmental monitoring and exchanges of information, CIS members have cooperated on such significant transboundary problems as air pollution, water supply and pollution, and the effects of the Chornobyl' accident. However, mutual suspicion and lack of funds often militate against closer agreement.

Conclusion

There is little evidence of rapid progress in dealing with the unhappy environmental heritage of the Soviet era, and the transition to the market is exacerbating some problems, while adding new ones. Too many

Russians are still living in environments which are injurious to their health. In some cases the impacts are disastrous. It is clear that it will take many years of investment and further reforms to improve, the situation radically. In the meantime, the situation in some parts of the country, especially in big cities and in the south of European Russia, could also have negative economic consequences and interfere with economic recovery. Unfortunately, attitudes adopted during the long years of Soviet power still hold sway and are hardly likely to change quickly, despite the officially adopted quest for a safer environment.

Seven

Population: Urban and Rural Life

In the second half of 1991, the Russian population began to decline for the first time since the Second World War. Newspapers were filled with articles warning of the eventual dying out of the Russian people and of the national weakness which seemed to be implied by the demographic statistics. The reasons for this demographic problem (or catastrophe, in the opinion of some) are complex, but the current transition through which Russia is going is obviously implicated (DaVanzo, 1996). Previous chapters have discussed the impact of the transition on the politics, economy and natural environment of the new Russia. This chapter will focus on its social impact. Having looked at the overall demographic situation, the chapter will examine urban and then rural life. Agriculture, which was only briefly mentioned in earlier chapters, deserves a more extended treatment and will be discussed in the section on rural life.

Demographic Patterns and Migration

The 1990s were by no means the first occasion in the twentieth century when Russia experienced population decline. In fact, the country has faced a number of demographic crises, the effects of which are still to be felt today. The years of the civil war (1918–21) witnessed the deaths of perhaps two million people. Many more died in the accompanying epidemics and the famine of the early 1920s. In the 1930s, enormous numbers died as a result of forced collectivization and in various famines and purges. The population of the USSR stood at 194.1 million in 1939 but after the Second World War in 1950 it is estimated to have been only 178.5 million, despite the annexation of new territories in the west in 1940–5. Wartime losses among the Soviet population have been put at 7.5 million in the military and 6–8 million civilians (Hosking, 1992,

p. 296). According to one estimate, if the Soviet Union had not endured such huge population losses in the twentieth century, its 1991 population would in all likelihood have been 150 million greater than the 290 million it actually was (Haub, 1994, p. 4).

Russia's population, which stood at 147.5 million in early 1997, bears many of the scars of these enormous demographic losses, especially those of the Second World War. Thus slow growth in the late 1980s can be ascribed to a shortage of women of child-bearing age at that period. This is the result of the reduced number of births in the 1960s. This in turn reflects the wartime deaths of people who might otherwise have borne children who would have added to the cohort of births two decades or so later. A second long-term effect of the Second World War was the shortage of men relative to women because of the greater casualty rate among the former. This was one reason why the Soviet government was so keen to encourage women to work. Nowadays, however, the shortage of males only becomes apparent in the older age groups above 65 years, where it is obscured by the differential death rates between the two sexes.

A feature which particularly worried the Soviet authorities was the failure of the population to grow as quickly as was deemed necessary if the economy was to succeed. As noted in chapter 4, the command economy had a voracious appetite for labour, and many sectors and regions suffered from shortages. As time went on, more and more regions of the USSR completed the demographic transition from high birth rates (death rates fell dramatically across the USSR after the Second World War) and high rates of natural increase to low birth rates and low rates of natural increase. In Russia's case as a whole, the crude birth rate fell from 23.2 births per 1,000 population in 1960 to 14.6 in 1970, and then rose slowly to 17.2 in 1986 to fall away again towards the end of the Soviet period (table 7.1). Meanwhile death rates, which had been very low by international standards in the 1950s and 1960s as a result of the many premature deaths in the Second World War, rose from 7.4 per 1,000 population in 1960 to 11.0 in 1980, and fluctuated about this level to the end of the Soviet period. The overall result for the rate of natural increase was a sharp fall from 1960 to 1980, followed by a slight rise to 1986 and then another sharp fall thereafter.

The Soviet authorities worried that slow population growth in Russia and other European republics would exacerbate the labour shortages in those regions and lead to an overall ageing of the population. The fact that slow growth was less of a problem in the southern republics, particularly in Central Asia, was little consolation, since Central Asians were mainly rural and traditional in their way of life and unwilling to move to Soviet industrial centres. Seemingly incapable of forcing the command

Table 7.1   Russia: birth rates, death rates and rates
of natural increase per 1,000 population, 1960–91

Year	Birth rates	Death rates	Natural increase
1960	23.2	7.4	15.8
1970	14.6	8.7	5.9
1980	15.9	11.0	4.9
1985	16.6	11.3	5.3
1986	17.2	10.4	6.8
1987	17.2	10.5	6.7
1988	16.0	10.7	5.3
1989	14.6	10.7	3.9
1990	13.4	11.2	2.2
1991	12.1	11.4	0.7

Source: *Rossiyskiy statisticheskiy yezhegodnik*
(1994, p. 43).

economy to use its labour force more efficiently, the authorities hoped to
be able to influence population growth itself by pursuing pro-natalist
policies. There is in fact some evidence that the slightly improved demographic situation in the early to mid-1980s was in part a response to such
policies (French, 1995a, p. 90). However, the factors influencing demographic behaviour are so complex and uncertain that it is very doubtful
if any government can modify them beyond a limited point. The trends
in Russian birth rates down to the late 1980s, which were briefly described above, seem to be accounted for by at least three factors in
addition to government policy. One of these is the shadow effects of the
Second World War, which, for example, helped to cause the decline in
birth rates experienced in the 1960s. A second is a factor helping to
reduce birth rates throughout the developed world: urban living and
rising living standards, which became apparent from the late 1950s,
helped to foster new priorities and styles of life in which having a big
family was less important than before. The third is much more to do with
the peculiarities of the Soviet system – in other words, the deficiencies
and privations of everyday life. Shortages of accommodation, lack of
services of all kinds including child care facilities, the fact that women
were generally obliged to work as well as carrying the burden of household chores and of looking after the children, poor health and other
factors all militated against the idea of having more than one, or at most
two, children, at least in Russia and the other more developed republics.
Instability in family life was reflected in divorce rates, which tended to
rise after 1970. The frequent use of abortion as the principal means of

Table 7.2 Russia: birth rates, death rates and rates
of natural increase per 1,000 population, 1991–6

Year	Birth rates	Death rates	Natural increase
1985	16.6	11.3	5.3
1991	12.1	11.4	0.7
1992	10.7	12.1	−1.5
1993	9.4	14.5	−5.1
1994	9.6	15.7	−6.1
1995	9.3	15.0	−5.7
1996	8.8	14.3	−5.5

Source: Rossiyskiy statisticheskiy yezhegodnik
(1996, p. 50).

controlling family size may also have harmed the health of many women
and reduced their reproductive capacity.

The trend in death rates over the latter part of the Soviet period was
similarly caused by complex factors. The shadow effects of the Second
World War were again important, but they gradually waned after about
1965. While improvements in medical care and diet probably had some
positive influence, environmental deterioration and the failure of the
health care system to cater to changing needs had the contrary effect.
The continued prevalence of unhealthy practices like smoking and the
excessive consumption of alcohol frustrated the goal of increasing life
expectancy to levels common in other industrialized countries.

The Russian population's ability to reproduce itself suffered a severe
decline in the late Soviet period and became negative after 1991, as we
have seen. But the context in which this was occurring changed dramatically with the collapse of the USSR. No longer was it a case of worrying
about shortages of labour in different sectors and regions. In conditions
of deindustrialization and mounting unemployment, the traditional
Soviet concern about low population growth acting as a brake on the
economy no longer applied in the same way. Rather, commentators now
focused on the political repercussions of a declining population and the
longer-term implications of an ageing workforce and of the greater
incidence of premature death.

The birth rate, which had stood at 16.6 per 1,000 population in 1985,
when Gorbachev came to power, had fallen to 12.1 by 1991 and reached
8.8 by 1996 (table 7.2). In the late 1980s, much of the decline could be
explained by a fall in the number of women of child-bearing age, but
the continued decline thereafter must be put down to other factors.
Research suggests that the economic and social problems of the later

Gorbachev and Yeltsin years have had a severe impact on demographic behaviour (Morvant, 1995). Declines in real income, rising unemployment and the general uncertainties of life under transition seem to have persuaded many people to postpone their family plans or even marriage altogether. Divorce rates have risen and there has been an increase in births outside wedlock. What is not clear is whether all this is a temporary phenomenon or whether it portends more permanent changes in lifestyle.

Even more dramatic than the fall in the birth rate has been the increase in mortality. The death rate stood at 11.3 per 1,000 population in 1985 and 11.4 in 1991, but had climbed to 15.7 by 1994. By 1996, it had fallen slightly to 14.3. Only part of this overall increase can be accounted for by the ageing of the population. Whereas life expectancy at birth stood at 70.1 years in 1986–7 (64.9 for men and 74.6 for women), by 1995 it was only 65 years (58 for men and 72 for women). Life expectancy for men is thus now at a level of some low income countries in the Third World, while the difference between men and women is quite remarkable. The statistics suggest that 'external' causes of death (accidents, homicides, poisonings etc.) have currently replaced cancer as the second most important after circulatory diseases (diseases of the heart and blood vessels). Among men, external causes are responsible for much of the recent increase in mortality. Thus violent crime, suicide, industrial accidents (related to increasingly poor safety standards, especially in the burgeoning second economy), road accidents (related to rising traffic levels and poor maintenance of roads and vehicles), alcoholism (no doubt related to stress but also to the consumption of substandard and frequently home-made alcoholic substances) and drugs have all taken their toll. Among women, such causes of death have been less significant, although also growing in absolute terms. More important has been the growing death rate among elderly women, which is believed to be related to general impoverishment. Other factors which seem to be influencing the recent rise in mortality include rises in infant mortality (beyond the purely statistical rise caused by adoption of the internationally accepted definition of infant mortality), a rise in maternal mortality, a general deterioration in health care (only 1.8 per cent of the state budget was spent on health care in 1994, compared with 3.4 per cent in the Soviet era), an increase in the incidence of infectious diseases, poor diet connected with general impoverishment and deterioration in public hygiene (contamination of food and drinking water). Environmental degradation, which has been cited by some as a major cause of the recent rise in mortality, can be only a partial explanation in view of the fact that environmental problems have been around for a long period (see chapter 6), while mortality has risen suddenly (Ellman, 1994).

Table 7.3 Birth, death and natural increase rates per 1,000 population by selected region, 1995

	Birth rates	Death rates	Natural increase	Migration growth (thousands)
Russia	9.3	15.0	−5.7	502.2
North	8.7	14.2	−5.5	−25.3
Karelia	8.5	16.3	−7.8	1.8
Northwest	7.2	17.3	−10.1	40.3
Central	7.7	17.3	−9.6	166.2
Volga–Vyatka	8.6	15.8	−7.2	31.6
Chuvash Rep.	10.2	13.0	−2.8	3.8
Central Black Earth	8.5	16.3	−7.8	62.6
Volga	9.3	14.1	−4.8	104.7
North Caucasus	12.0	13.6	−1.6	86.4
Dagestan	21.8	7.5	14.3	0.4
Ingushetia	23.8	6.4	17.4	14.3
Urals	9.5	14.5	−5.0	74.4
Bashkortostan	11.2	12.7	−1.5	22.5
West Siberia	9.4	13.5	−4.1	49.7
Yamalo-Nenets AO	13.1	6.4	6.7	5.1
East Siberia	11.0	13.7	−2.7	3.9
Tyva	20	13.0	7.0	−0.5
Far East	10.2	12.6	−2.4	−102.8
Sakha Rep.	15.3	9.8	5.5	−18.7

Source: *Rossiyskiy statisticheskiy yezhegodnik* (1996, pp. 721–3, 733–5).

Locally, however, poor environmental conditions can be an important cause of ill health and mortality: witness the nickel and copper smelting centre of Nikel' on the Kola Peninsula, where life expectancy at birth was only 34 years in 1992, or the Noril'sk metallurgical complex in northern Siberia, with a cancer rate of 485 cases per 100,000 population in 1989 compared with a national rate of only 268.

The geography of birth, death and natural increase rates across the Russian Federation is a complex one and not easily explained. The indicators are, however, generally more positive, with higher birth and natural increase rates, in some peripheral territories (Siberia, the Far East) and especially in autonomous regions. The populations of such regions tend to be younger than the average (table 7.3).

In summary, then, the post-communist transition has had a major impact on population growth patterns, with many negative implications for health and social welfare. A related area where the transition has also had a major impact is migration. As is the case with demographic behaviour generally, interregional migration since 1991 displays a strengthening of tendencies which were already becoming apparent towards the end of the Soviet period. Although some scholars have made reference to the 'quasi-military powers' available to Soviet planners in controlling population movements (Cole and Filatochev, 1992, p. 433), the Soviets were not in fact particularly successful in persuading the population to live where they were needed. During and after the Second World War, direction of labour was used and people were forbidden to move jobs without permission. But this was abandoned in the mid-1950s as inefficient, and greater reliance had to be placed on financial and other kinds of incentive, including moral incentives where appropriate. The Soviet authorities retained in their hands certain other instruments for controlling migration, such as compulsory placements for new graduates and a system of internal passports and residence permits (the latter particularly for big cities and certain sensitive locations), but it is doubtful if these had a big impact on overall migration patterns.

The major aim of Soviet migration policy was to induce people to locate in areas of labour shortage, especially in the resource frontier regions of the north and the east (figure 7.1). Although this was reasonably successful in the case of certain high priority projects – the Stalinist drive to develop eastern resources (accompanied by much forced labour), the Virgin Lands campaign of the 1950s (when large areas of semi-arid terrain in West Siberia and northern Kazakhstan were put to the plough), the development of the north-west Siberian oil and gas fields from the late 1960s and the Baykal–Amur Railway project in the 1970s – it was always difficult, especially in the frontier regions, to create incentives which were powerful enough to persuade people to stay permanently. A high turnover of labour was therefore the norm. In the 1960s, after many of the Stalinist regulations were abolished, numerous eastern regions suffered net outmigration. Later the situation improved somewhat as a result of better incentives, but towards the end of the Soviet period outmigration was rising again as a result of the difficulties and changed priorities of *perestroyka*. During 1986–90, the North, Urals and East Siberia all suffered from net outmigration, while the net inmigration into West Siberia and the Far East was much below what it had been in the previous period (Gibson, 1994). Since 1991, the situation for many northern and eastern regions has been little better than catastrophic. The withdrawal of many government subsidies, rising transportation tariffs, the reduction of the military presence, the closure of some economic activities and the desire of some citizens of the other newly independent states to return home have precipitated considerable outmigrations. Murmansk Oblast in north European Russia suffered a 10.2 per cent decrease in its population between 1989 and 1996, the

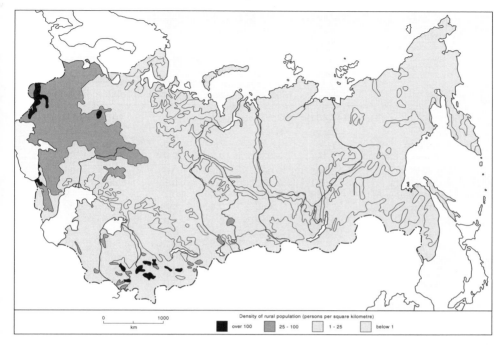

Figure 7.1   Former USSR: distribution of population.

Komi Republic 7 per cent, while in the Far East, Kamchatka Oblast's population fell by 13.6 per cent, and Chukchi Autonomous Okrug's by 44.3 per cent (table 7.4).

A further characteristic of Soviet migration patterns, though one which was not always a direct result of policy, was the tendency of Russians in particular to move into other Soviet republics. As noted in an earlier chapter, some 25 million Russians lived outside the Russian Federation at the end of the Soviet era. Much of this was the result of migrations which had taken place since 1917, as Russians had moved to take up industrial, military, service and administrative positions in the other republics or sometimes (as in the Baltic states) to benefit from higher living standards. Down to the mid-1970s, Russia was a net exporter of migrants. After that time, however, Russia became a net importer, as more and more Russians migrated back to their home republic, partly in the wake of growing national consciousness and economic competition in the other republics. Since the fall of the USSR this movement has become a flood. Between 1989 and 1996, Russia experienced a net inmigration from the other post-Soviet republics of

Table 7.4   Migration trends in selected Russian regions, 1989–96 (in thousands)

Region	Total population		Percentage change 1989–96		Absolute change 1989–96	
	1989	1996	Total	Migration	Total	Migration
Russia	147,401	147,501	0.1	1.7	101	2,510
North	6,123	5,833	−4.7	−3.9	−290	−241
Komi Rep.	1,261	1,172	−7.0	−8.0	−89	−101
Murmansk Obl.	1,147	1,030	−10.2	−10.9	−117	−125
West Siberia	15,003	15,087	0.6	0.8	84	123
Tyumen' Obl.	3,081	3,177	3.1	−0.6	96	−18
Khanty-Mansi AO	1,268	1,331	5.0	−0.7	63	−9
Yamal-Nenets AO	486	488	0.4	−6.3	2	−30
East Siberia	9,155	9,112	−0.5	−1.6	−43	−151
Tyva Rep.	309	310	0.4	−8.5	1	−26
Taymyr AO	55	47	−15.5	−19.2	−9	−11
Evenki AO	24	20	−15.4	−20.5	−4	−5
Far East	7,941	7,421	−6.5	−8.1	−520	−646
Sakha Rep.	1,081	1,016	−6.0	−13.0	−65	−141
Chukchi AO	157	87	−44.3	−48.3	−70	−76
Kamchatka Obl.	466	403	−13.6	−15.3	−64	−71
Magadan Obl.	386	251	−34.9	−36.8	−135	−142

Source: Heleniak (1996, pp. 86–8).

about 2.8 million people. At the same time, the country experienced a net outmigration to the rest of the world of about 600,000 people (Heleniak, 1995, 1997). Some of the factors which have led people to move from the other post-Soviet republics to Russia will be discussed in chapter 10. Here the point needs to be made that a portion of the immigrants have been officially classified as 'refugees' (those not qualifying as citizens of Russia) and 'forced migrants' (those qualifying). In the circumstances attending the break-up of the USSR, however, it is very difficult to distinguish meaningfully between forced and voluntary migration. Even so, it is clear that the migration of such numbers into Russia is putting great pressure on housing, services and other facilities, especially when the country is trying to cope with military personnel displaced from the other republics as well as from the former Warsaw Pact countries. Recently, the Russian authorities have been trying to limit immigration. A number of regions have been designated as reception regions for migrants, and there are schemes to settle some of them in needy agricultural areas. However, it is very doubtful whether many regions are equipped to cope with the immigrants, or many immigrants with the agricultural life. In fact, rather than moving to the official reception areas, migrants seem to have been travelling to the bigger cities of central and south European Russia, where they evidently feel conditions to be better.

Urban Life

Throughout the Soviet period, people migrated from the countryside to the town in search of a better life. Stalin's exploitative attitude towards the peasants was sufficient to impel many towards the cities. But both in Stalin's day and later the cities proved attractive in themselves, promising higher living standards and advantages such as jobs, low rents and welfare benefits unavailable in the countryside. Yet Stalin's policies engendered many problems for the cities. Early Bolshevik attitudes had tended to favour cities as the habitat of the working class. Urbanization was thus welcomed as fostering the proletarianization of society. Cities were the 'control points' of the regime, the places from which the political supervision of the countryside was organized. But they were also points of vulnerability, since they were the homes of large numbers of people who might rebel if urban problems proved too severe. The Bolsheviks were unlikely to forget that 1917 started as an urban revolution. However much Stalin may have wished to forget about the cities and concentrate totally upon the task of industrialization, he could not afford to do so. His answer was to try to control immigration into the biggest cities through a system of internal passports and residence permits, and to develop some basis for city planning (Parkins, 1953; French, 1995b). Great reliance was placed on the industrial ministries to provide housing, services and other necessities for their workers. But the massive rural-to-urban migration, inadequate investment and wartime destruction meant that the urban environment deteriorated none the less. The product of the Stalin years was a city which lacked many of the necessities of urban life, narrowly focused around the industrial and military priorities of the regime.

In the years after Stalin, a greater effort was made to raise living standards and solve pressing urban problems. Characteristically, the regime which had scored such notable successes on the industrial front now began the attempt to solve the problems of the cities using the same command economy approach. The huge resources of the industrial and construction ministries were focused on the mass production of urban housing utilizing Fordist industrial techniques. The result was a dramatic quantitative improvement in urban housing supply as cities were surrounded by massive suburbs of high rise, prefabricated housing (table 7.5). Attention also began to focus on other urban problems, such

Table 7.5 Development of the Russian housing fund (millions of square metres)

	1970	1980	1990	1991	1992	1993	1994	1995
Total	–	1,861	2,425	2,449	2,492	2,546	2,608	2,649
Urban	914	1,291	1,720	1,749	1,779	1,836	1,882	1,915
Private	240	304	353	356	418	634	726	797
Citizen-owned	217	244	260	261	322	481	576	646
State	420	602	765	773	738	401	277	209
Municipal	254	382	594	612	611	638	697	742
Social	–	3	8	8	12	2	2	1
Joint	–	–	–	–	–	161	180	166
Rural	–	570	705	700	713	710	726	734
Private	–	399	438	446	468	555	587	601
Citizen-owned	–	376	381	390	428	462	502	520
State	–	169	246	232	211	95	72	61
Municipal	–	–	17	17	19	26	35	45
Social	–	2	4	5	15	1	1	1
Joint	–	–	–	–	–	33	31	26

Source: *Rossiyskiy statisticheskiy yezhegodnik* (1996, p. 236).

as retail and service provision, and environmental planning. And since the attempt to solve all these problems raised many questions concerning the overall organization and management of cities, it was necessary to revamp the urban planning machinery. In every case, the same reliance was placed on central control, standardization and mass production methods (Bater, 1980, pp. 97ff.).

In the event, policies to improve the quality of life in Soviet cities proved inadequate, and their inadequacies became more and more apparent as time went by. The Soviet economy failed to generate sufficient investment for the cities, and, as growth rates began to slow down from the 1970s, the shortfall became more obvious. The planning system proved unequal to the task of controlling urbanization in an increasingly complex society. And the attempt to solve urban problems by using the methods of the command economy – mass production and standardization, sometimes described as the 'extensive' approach to urbanization – was unwieldy, inflexible and insensitive to the needs of the populace.

Russia has experienced a very rapid rate of urbanization since 1917, and today nearly three-quarters of the population live in urban places (table 7.6). Despite the system of residence permits and internal passports which operated in the Soviet period (and to a limited extent today), much of the urbanization was spontaneous and outside the control of the authorities. The government in fact had limited control over the evolution either of individual towns and cities or of the settlement system

Table 7.6 Russia: total and urban population,
1917–96 (millions)

Year	Total population	Urban population	% urban
1917	91.0	15.5	17
1939	108.4	36.3	33
1959	117.5	61.6	52
1970	130.1	81.0	62
1979	137.6	95.4	69
1989	147.4	108.4	74
1990	148.0	109.2	74
1991	148.5	109.8	74
1992	148.7	109.7	74
1993	148.7	108.9	73
1994	148.4	108.5	73
1995	148.3	108.3	73
1996	148.0	108.1	73

Source: Rossiyskiy statisticheskiy yezhegodnik
(1994, p. 17), ibid. (1996, p. 37).

as a whole. One consequence, and an undesirable one in what was
supposed to be a socialist society, was a considerable unevenness in the
level of urbanization across Soviet and Russian territory (table 7.7).
Urbanization went hand in hand with industrialization, leading to the
appearance of towns even in quite remote locations, sometimes with only
a single industry to sustain them. From the 1970s (and especially in the
1990s), there has arisen a category of demographically declining towns
in consequence of the contraction of their economic functions
(Rowland, 1994, 1995, 1996a). Monofunctional or narrowly specialized
towns formed a significant proportion of such declining towns.

A related feature of the Soviet mode of urbanization was the enhanced
importance of big cities. In 1996, Russia had a total of 168 cities with
over 100,000 population, containing 62 per cent of the urban population, or just over 45 per cent of the total population (table 7.8). Some
four-fifths of these big cities are located in the European part of the
country. Their importance in Russia's urban geography relates to the
fact that it was (and is) the big cities which were endowed with the best
infrastructures and quality of life, making them attractive to economic
organizations and migrants alike. The residence permit system proved
incapable of controlling their growth. A notable feature of the latter part
of the Soviet era was the tendency for the urban population of many of
Russia's regions to concentrate in the region's biggest and generally most
favoured city, namely the administrative centre.

Table 7.7   Levels of urbanization across Russian economic regions, 1996 (percentages)

Russia	73.0
North	75.7
Northwest	86.7
Central	82.9
Volga-Vyatka	70.1
Central Black Earth	61.6
Volga	73.0
North Caucasus	55.6
Urals	74.5
West Siberia	71.0
East Siberia	71.4
Far East	75.8

Source: *Rossiyskiy statisticheskiy yezhegodnik* (1996, pp. 711–13).

Table 7.8   Size distribution of Russian urban settlements, 1996

Size distribution (thousands)	Number of urban settlements	Total population (thousands)
All urban settlements	3,109	107,671
Below 3.0	593	1,033
3.0–4.9	468	1,843
5.0–9.9	741	5,285
10.0–19.9	555	7,752
20.0–49.9	407	12,854
50.0–99.9	177	12,170
100.0–499.9	137	29,926
500.0–999.9	19	12,603
1 million and above	12	24,205

Source: *Rossiyskiy statisticheskiy yezhegodnik* (1996, pp. 68–9).

Entirely different is life in Russia's 2,750 or so small towns with fewer than 50,000 people. These towns contained 27 per cent of the urban population in 1996. Many are characterized by very poor levels of servicing and low living standards, and often by narrow economic profiles. When such towns are located in oblasts with only one big city, the administrative capital, and an absence of intermediate centres, they tend to have a particularly poor quality of life. This is the case in some rural regions of central and southern Russia, and in parts of the east.

Table 7.9 Level of urban development of Russian economic regions, 1989.

	Level of urban development (LUD) (%)	Official percentage of population urban (PPU)	Excess of PPU over LUD (% points)
Russian Federation	40	66	26
North	36	77	41
Northwest	69	87	18
Central	55	83	28
Volga–Vyatka	40	69	29
Central Black Earth	35	60	25
Volga	53	73	20
North Caucasus	30	57	27
Urals	47	75	28
West Siberia	47	73	26
East Siberia	40	72	32
Far East	38	76	38

Source: Pivovarov (1992, p. 55).

It is on the grounds of the unsatisfactory quality of urban life in many Russian towns (mainly small, but also some with populations exceeding 100,000) that some Russian geographers have argued that the real level of urbanization in Russia is less than the official statistics suggest (Pivovarov, 1992). Pivovarov has recalculated the 'real' urbanization level by estimating the extent to which the population engages in urban-type occupations and activities, enjoys an urban standard of services, benefits from a genuinely urban environment and is exposed to urban culture. He and others have argued that 'pseudo-urbanization' (i.e. over-urbanization, a characteristic feature of parts of the Third World) is particularly apparent in Central Asia but also in other parts of the former Soviet Union, including some regions of Russia (table 7.9). According to his recalculation, the most urbanized regions of Russia are the Northwest, the Centre and the Volga, and the least are the North Caucasus, the Central Black Earth Region, the North and the Far East. Particularly wide discrepancies between the official and the 'real' level of urbanization are characteristic of the North, the Far East and East Siberia.

Pivovarov's arguments seem opposed to those of some other scholars, who maintain that the former USSR is a case of 'under-urbanization' (Ofer, 1976). This is to say that the former USSR was less urbanized than it should have been judging by its degree of industrial development:

many big cities would have been even bigger had it not been for the residence permit system and accommodation shortages which forced part of their labour forces to live in the surrounding region and commute in. Other scholars have compared the evolution of the former USSR's employment structure with that of other countries and concluded that the USSR's urbanization levels probably lagged somewhat behind the changing employment patterns (Treyvish et al., 1993). Clearly, different scholars have different definitions of 'urbanization'. But the significant point is that all are agreed that the Soviet urban system was seriously underdeveloped.

Soviet urbanization, then, promoted regional discrepancies, with some regions having considerable numbers of big cities (sometimes forming urban agglomerations) and others characterized by a predominance of small and poorly developed towns. There were also wide discrepancies between urban places even in the same region, with big cities enjoying a more developed lifestyle than smaller places, to say nothing of rural settlements. The characteristic Soviet pattern of development by means of the branch industrial ministries did nothing to promote harmonious regional planning or interrelationships between towns, cities and other settlements, even when they were situated side by side. Each ministry preferred to concentrate on its 'own' settlements or urban districts. In pioneering regions such as parts of Siberia and the north, industrial ministries often owned entire towns whose life revolved around perhaps one large enterprise. Such 'factory' towns were notorious for their poorly developed infrastructures and service levels, the ministries typically having other priorities. In larger cities too, ministries might develop their own 'villages' close by their enterprises, with housing and services provided for their employees. While those fortunate enough to work for 'good' ministries might thus enjoy living standards above the average, the system promoted spatial inequality and made it difficult for city governments (soviets) to plan the city as a unity. Despite attempts by city soviets to coordinate the urban investments of the ministries, the latter frequently failed in practice to contribute their share to urban development. Other disadvantages also flowed from the accent on industrial development. Industrial enterprises often occupied huge tracts of land (over 30 per cent of that in Moscow, for example, compared with only 5 per cent in Paris – land had no market price under Soviet conditions), and might intermingle with housing or push housing development to the periphery, thus adding to commuting distances and to pressure on public transport. Land use planning provisions were often violated by powerful ministries. The detrimental effects of industrial development on the environment have been considered in chapter 6.

The Russian city-dweller of today must therefore cope with numerous problems bequeathed by the Soviet era. The narrow, industrially oriented approach to urban development meant that many essential elements of the city – housing, services, public transport, many infrastructural features – were neglected. In housing, for example, where public housing predominated, it was common for families to wait up to ten years for a suitable apartment. Married couples often lived with parents in cramped conditions. This did nothing to raise birth rates, needless to say. Divorcees were also frequently obliged to live in the same apartment, adding to the stresses and strains of daily life. In shopping, continual shortages of goods of all kinds (to say nothing of the prevailing poor quality) often meant people having to visit numerous shops to secure the items they required. Furthermore, the antiquated system of shopping, which was itself a product of shortages and the need to combat dishonesty, meant it was often necessary to queue several times just to buy one item. The overcrowded public transport system was yet another feature making the Soviet city a stressful environment for living. Monotonous townscapes (the product of the industrialized approach to urban development), inadequate provision of open space and a poor environment generally did nothing to ease the other shortcomings of the city. Unfortunately, given the highly centralized and autocratic character of the Soviet political system, neither individuals nor city soviets had the resources or the power to address the problems of the city. Such problems are unlikely to disappear quickly in the new Russia.

The individual's circumstances in the new, post-Soviet city will depend upon two major factors. The first is the position of the city in which he or she dwells in relation to the major economic changes which were discussed in chapter 5. While deindustrialization is affecting every region, 'gateways', 'hubs' and some cities in resource-producing regions will be far better placed to benefit from the new economic changes than those located in declining industrial areas or the peripheral, agricultural ones. Conditions are likely to be particularly bad in monofunctional towns where the basic economic activity is in decline. Barring government intervention in some form (for example, to bolster the military–industrial complex or to sustain the existence of the 'closed' cities; see Rowland, 1996b), there will be little or no money available for urban improvement. Successful cities in economically dynamic regions, or those less well placed which nevertheless manage to sell themselves to domestic and foreign investors, will have the money to invest in urban services and environmental improvement. Having the means to do such things is now more important than ever, as regional and city authorities have been forced to take over ever more responsibility for meeting the needs of their citizens from a cash-starved central government. Likewise for the individual, employment in the city's increasingly privatized economy, supplemented perhaps by earnings from the burgeoning shadow economy, will provide the means to gain access to services which were hitherto provided free of charge by the state. The city's economic success will also be affected by non-economic factors. Political instability and/or ethnic strife will obviously have a detrimental effect. If the city becomes a target for large numbers of refugees or interregional migrants, this might also complicate matters, depending on how the city copes. Similarly, high crime rates might deter investors, though the example of Moscow suggests that crime is not necessarily incompatible with economic growth. And what of the environment? A poor environmental record might damage the city's prospects, as will a 'rustbelt' image. Equally, however, economic success can bring its own environmental problems – one has only to think of the growing automobile pollution in Moscow, St Petersburg and other cities.

The other major factor to affect individual circumstances is the individual's position in the emerging social structure. This partly depends on the heritage of the past. The point has already been made that the Soviet system begat social inequality. Such inequality was reflected in the social geography of the city (Bater, 1986; French, 1987). The most privileged members of Soviet society, known as the *nomenklatura*, who held important positions in party and state, plus a handful of others such as top artistes and sportspeople, often had prestige apartments close to the shops and services of the city centre (typically rented to them at very low rents by their state employers), and perhaps a dacha for weekends in the countryside. What mattered in Soviet society was not so much what such people earned, though that was generally well above the norm, but the fact that they usually enjoyed privileged access to housing, shops, health care, tourist facilities and other services denied to the majority of the population. As far as the majority of the population was concerned, much depended on the jobs they occupied (with the exception of some parts of Central Asia, unemployment was not a problem in Soviet society). The better off might have enough money to buy a cooperative apartment, a car and other luxuries. If they worked for a 'good' ministry, they too might enjoy superior accommodation and services. Most people were forced to rely on their state employers for accommodation, or alternatively on the city soviet. The problem here was not the cost of accommodation, since rents were universally low, but acquiring it in the first place and the poor maintenance of many apartment blocks. Location might also be a problem, as noted above – the mass housing built from the late 1950s tended to be in the suburbs, miles from the city centre, services and even workplaces in big cities. Those who were not lucky enough to have a state apartment might live in private housing (about 20 per cent of urban housing but more in smaller towns). This was generally of poor quality. Alternatives were to live in a communal apartment with other families (usually in older buildings) or sub-let space in someone else's flat. The least well off were those living in hostels (usually young and sometimes on limited residence permits), or long-distance commuters who travelled into the city by bus or train, generally to poorly paid jobs.

Although the Soviet system gave rise to inequality, there were certain features of that society, like guaranteed employment, low rents and state welfare, which prevented most people from descending into abject poverty. The same seems not to be the case in post-1991 Russia, where a new order of social inequality seems to be in the making. At one end of the scale are the new entrepreneurs (the 'new Russians'), who have made money from the growing capitalist economy (including organized crime in some cases), and who sometimes flaunt their wealth in the crudest possible way. Also wealthy are many members of the former *nomenklatura* who have found ways of profiting from policies of privatization. At the other end of the scale are the large numbers of poor and destitute now to be found on the streets of numerous cities, the victims of inflation, unemployment and physical disability (including old age). The majority of people cope as best they can: continuing to work for their former employers where this is possible and desirable, seeking work in the new capitalist businesses, supplementing their incomes in the second economy, taking to crime and so on. The new Russia is a harsher and more competitive place than the old, yet this encourages forms of cooperation between family and friends. For example, the growing of food on one's own allotment has become an important means of subsistence for many families. In the opinion of some, this represents a ruralization of the city, a partial return to the peasant ways which are still part of the folk memory of many urban Russians.

The privatization of housing is the means whereby the state has been passing on the burden of building and maintaining housing to the householders. In many cities this has taken the form of a virtually free acquisition of assets by existing tenants. The housing patterns of the old regime have thus been preserved, to the benefit of the already privileged. It remains to be seen whether the least privileged under the old system will now benefit from a new programme of social housing construction. Thus far, the omens seem hardly promising. In the meantime, the market in property and land is only beginning to develop (Gdaniec, 1997). There seems little doubt, however, that in future the city centre will become less of a residential quarter than in the past, as new retail outlets, offices, hotels and other functions barely provided for in the

Soviet city begin to appear. No doubt the suburbs will also be better serviced, particularly the less industrial, or socially and environmentally more desirable ones. The construction of new elite estates of low rise housing, a process which has already begun in many cities, will also enhance social differentiation in the city, as will the gentrification of some inner-city districts.

A telling symptom of the problems being encountered in big Russian cities at the present time is the fact that many of them having been losing population. Rowland records 49 declining cities over the period 1989–93 with populations in excess of 100,000 (Rowland, 1995). Interestingly enough, the list included the six biggest cities in Russia: Moscow, St Petersburg, Nizhniy Novgorod, Novosibirsk, Yekaterinburg and Samara. All these were still declining as of 1996. Big cities have long suffered from low birth rates and have relied on inmigration for continued growth. In the most recent period, not only have birth rates fallen even further, but outmigration has overtaken inmigration in many cases. The distresses of the transition period, including job losses and declines in the quality of life, appear to be implicated. This trend thus represents a reversal of patterns predominating in the Soviet period, when big cities proved the most powerful magnets for migrants. What is less certain is whether this is a temporary phenomenon, or whether it is the first phase of the counter-urbanization which has characterized numerous developed countries.

It will clearly take many years to solve the urban problems inherited from the Soviet era. Yet the cities are already changing, some of them quite quickly, with the new offices, retail outlets, banks, elite housing, street corner kiosks and markets acting as harbingers of capitalism. Nowhere are such changes reflected more vividly than in the capital, Moscow, whose 850th anniversary celebrations in 1997 became a major exercise in civic boosterism. Whether such developments will yet benefit the lives of the majority of Russian city dwellers remains a disturbing question for the future.

Rural Life and Agriculture

Russia is now a highly urbanized country, but in the mid-1990s more than a quarter of its population (some 40 million people) still lived in the countryside. Agriculture and forestry constitute the occupation of about 15.7 per cent of the employed population. The impact of the current changes on the countryside therefore constitutes an important part of the entire picture, and agricultural performance remains of key significance for the country's economic future.

Sadly, Russia's heritage in the countryside is even less promising than that in the towns. Stalin's forced collectivization, which did such damage in the early 1930s, meant the replacement of the traditional peasant landholdings by a system of large collective (and growing numbers of state) farms to which the peasants became compulsorily bound – almost a return to serfdom. Collectivization allowed the government to control the distribution of the farm's product, something which had been a real bone of contention in the 1920s. But coupled with the neglect of agricultural investment, the system became a truly exploitative one, reducing the rural population to a level barely above slavery. Under Stalin and later, the peasants found little incentive to show initiative or to work hard. Inheriting their peasant status from their parents, they were denied the automatic right to leave their farms and for long their labour was miserably underpaid. One redeeming feature was the personal plot of land, usually adjacent to the house, which peasants on collective farms were allowed to retain from the 1930s, initially for subsistence purposes. Because such plots became extremely productive, the peasants won the right to sell their surpluses to their farms or even on the restricted open markets. Thus this private sector, accounting for a tiny percentage of the land, came to produce a significant proportion of such high-value products as potatoes, vegetables, eggs, milk, meat and wool, albeit in close symbiosis with the collective and state farms (Lerman et al., 1994).

Given the low status of the peasants and the lack of incentives, it is hardly surprising that Soviet agricultural performance was poor. Under the command economy, agriculture was managed in much the same way as industry. Plans were passed down the bureaucratic hierarchy to the collective and state farms and the latter produced in response to orders from above. Yet agriculture has a number of characteristics which distinguish it from industry. A key one is the very significant and partially unpredictable influence of the natural environment, particularly the weather. Another is the need for detailed local knowledge if agricultural activities are to be conducted efficiently. Central planners found it impossible to have the detailed and up-to-date knowledge which agricultural decision-making requires, and the system militated against flexibility. For this reason, great reliance was placed upon local agricultural administrators at district level, including Communist Party officials, although this is very different from relying on practical farmers. Moreover, given the historic communist tendency to treat the peasants with suspicion if not disdain, local officials usually found the temptation to interfere in the day-to-day running of farms irresistible. Officious interference by bureaucrats undermined the self-confidence of farm managements and the self-respect of the ordinary peasants. Equally, however, trying to increase peasant incentives in various ways, a policy pursued in the post-Stalin period, proved only partially successful. One reason may have been that payments made to farms bore little relationship to efficiency (Cook, 1992, pp. 204–5). The effectiveness of other policies of the period, such as land improvement and investment in machinery and agro-chemicals, was also disappointing and proved hugely expensive to the state. Inadequate inputs, and the broader economic environment within which agriculture had to operate, must share part of the blame (Cook, 1992).

Although rural life was always viewed by the Soviets in the context of agriculture, they were aware that policies to make the latter more productive were unlikely to succeed without paying attention to the quality of rural life generally, including living standards. From the 1950s, therefore, some attention was paid to the need to improve rural life by investing in roads, services, better housing and other facilities. Given the extremely backward state of much of the rural infrastructure, a backwardness dating from tsarist times but exacerbated by Stalin's policies, the need was enormous and even today much remains to be done. An important feature of post-Stalin policy was the attempt to concentrate resources on key centres and villages, leaving the others to atrophy or be officially effaced. As Pallot has shown, all too often this policy was applied in official ignorance of local conditions (Pallot, 1979, 1990). Not only did its mechanical and rigid implementation have negative consequences for those living outside key villages, but the imposition of planning norms derived from urban experience made scant allowance for the special needs of the key villages themselves. Such policies did nothing to counter dissatisfaction in the countryside or to combat the alternative attractions of the city. It is therefore hardly surprising that many peasants 'voted with their feet' and abandoned the countryside for the town. Rural depopulation on an alarming scale was thus a characteristic feature of many parts of the European USSR as well as of Soviet Russia's eastern territories.

Of course, agricultural underperformance, and rural poverty generally, are not entirely the fault of Soviet policy. The natural environment is also a problem. Some of the difficulties were discussed in general terms in chapter 1. Natural conditions mean that less than 8 per cent of the vast territory is used for arable farming (about 10 per cent in the old USSR), and even within this region environmental conditions are often poor (figure 7.2). N. C. Field compared the former USSR's cropland resources with those of North America, and came to the conclusion that environmental quality plays a significant role in explaining the differences in agricultural productivity between the two regions (Field, 1968). In terms of both thermal resources and the availability of moisture, the

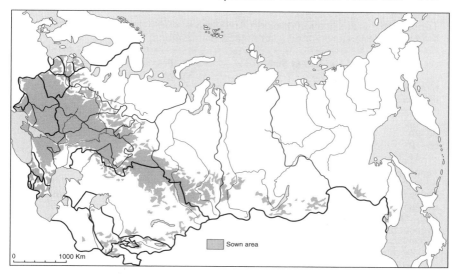

Figure 7.2   Former USSR: distribution of arable land.

former USSR suffers from having a far higher proportion of its cropland in regions with unfavourable characteristics than is the case in North America, especially the USA (table 7.10). Some significance must also be attached to land degradation and the loss of land to non-agricultural uses, both, of course, basically the result of human action. This is an issue which was briefly discussed in chapter 6.

Natural environmental conditions are taken into account in the Russian Academy of Sciences Institute of Geography's regionalization scheme for the former European USSR (with the exception of the Transcaucasus), based upon indicators of agricultural intensity (Ioffe et al., 1989) (figure 7.3). What this shows is that the most intensive and productive agriculture is found along the western and southern fringes of the former USSR. The area defined as the northwestern region is the most intensive, corresponding with the three Baltic states, Leningrad Oblast, the Karelian Republic and Moscow Oblast. A proportion of this region, therefore, falls outside the Russian Federation. The other two regions of reasonably intensive agriculture, the western (corresponding with the western parts of Ukraine, Belarus' and also Moldova) and the southern (the rest of Ukraine and the North Caucasus economic region) suffered from underinvestment, namely in labour productivity in the former case, and in both labour and land productivity in the latter. Of these regions, only the North Caucasus falls in Russia. The remaining

Table 7.10   Classification of North American and former Soviet cropland by thermal and moisture zones (percentage distributions)

Moisture classes[b]	Canada and United States Degree-months[a]				Former USSR Degree-months[a]			
	100–199	200–299	300+	Total	100–199	200–299	300+	Total
90–100	10	25	15	50	26	0.1	0.3	26
80–89	4	4	2	10	14	1		15
65–79	13	5	3	21	18	6		24
0–64	7	5	7	19	22	9	4	35
Total	34	39	27	100	80	16	4	100

Notes: [a] Thermal classes defined on the basis of accumulated summer temperatures measured in degree-months (the sum of mean monthly temperatures in excess of 0°C). [b] Moisture classes defined in terms of the percentage of water need, or potential evapotranspiration, which is satisfied through actual evapotranspiration. Higher percentages signify more humid conditions.
Source: Field (1968, p. 10).

three agricultural regions, namely the eastern, the northern and the central, all fall in Russian territory, with the exception of part of the central area, which corresponds with eastern Belarus'. All were regions suffering from various shortfalls in investment, and the first two have natural environmental problems. The central area, a huge region which embraces both the Central Black Earth and parts of the non-black earth regions, had particularly complex problems. These included low agricultural productivity, widespread outmigration and a contracting agricultural area. The southern parts, however, have rather favourable environmental conditions.

The break-up of the USSR has therefore left Russia without much of the most productive agricultural territory in the former European USSR (where, it will be remembered, most of the USSR's agricultural land was located). Considerable investments would be needed to develop the agricultural lands still falling within Russian territory. An important finding of the Institute of Geography's study is that there was little obvious correlation between investment patterns and the potential for increasing agricultural output. There existed a wide zone in the central part of European Russia where yields fell below even what might have been expected under natural moisture conditions. In other words, the irrationalities of the command economy greatly compounded the difficulties produced by nature.

In the 1980s, the Soviet Union found it necessary to import more than 20 per cent of the calorie content of the population's diet, and agriculture was taking up to a quarter of total investment. This was a

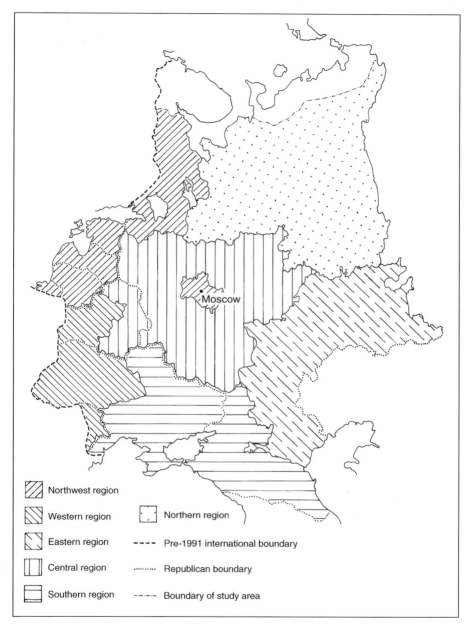

Figure 7.3 European USSR: agricultural zonation.
Source: after Ioffe et al. (1989).

Legend:

- Northwest region
- Western region
- Eastern region
- Central region
- Southern region
- Northern region
- ---- Pre-1991 international boundary
- .... Republican boundary
- ---·--- Boundary of study area

Moscow

huge drain on resources which were badly needed for other purposes. Therefore, both Gorbachev and Yeltsin decided to try to tackle agriculture's problems head on. Given the general orientation of their economic reforms towards decentralization and marketization, it was probably inevitable that similar policies should be applied in the countryside. The argument was that collectivization had broken the historic link between the peasants and their land, thus undermining incentives. However, rather than an attempt to return to traditional Russian farming, in which peasants had tended to farm their land in the context of communal institutions (Bartlett, 1990), the model which appealed most strongly to reformers was the privately owned, commercially oriented family farm which has been such a characteristic feature of western Europe and North America. The argument was that these farms were the type most appropriate for the newly marketizing economy, and that they were much more likely to achieve commercial success than the old collectives. Russia's only previous experience of such farms was during the Stolypin agricultural reforms in the early years of the twentieth century, but these had disappeared after the 1917 revolution.

The first move in the direction of privatizing the Soviet countryside came with policies to allow groups and then individual families to resume responsibility under contract for specific activities or areas on the farm. At the same time, other measures gave further encouragement to the personal plot and to collective gardening by city dwellers. In 1988, Gorbachev decided to permit land and property leasing by collective and state farms for the establishment of individual farms. The lessees could be individuals or cooperatives. Then, in June 1990, farmers were given the right to secede from their collective and state farms with their share of the land. These were radical measures by communist standards, but the Soviet leadership was careful to hedge its reforms with restrictions, in deference to more conservative opinion.

In the meantime, however, agriculture began to suffer severely as a result of other measures implemented by the Gorbachev leadership. Thus farms were obliged to assume responsibility for their own financial affairs, but marketization of prices was only partial and the growing discrepancy between the price of inputs and outputs, and the increased demand from consumers because of wage inflation, led to food shortages. Agriculture was not immune to the supply problems which afflicted the Soviet economy under Gorbachev, the result of growing dislocation on all sides (see chapter 4). The difficulties were exacerbated by Gorbachev's attack on the bureaucracy. Abolition of ministries and other state agencies, and constant reorganizations, disrupted links between farms, their suppliers and their procurement agencies, with no real market to fill the vacuum. The result in some regions was a return to traditional methods of bartering. Regional and local authorities often intervened with all kinds of restrictions and regulations in an attempt to control the food supply.

Agriculture was one of the arenas chosen by Yeltsin for his struggle with the dying Soviet government, and even before the final collapse of the latter the Russian leadership was trying to introduce more radical measures in the countryside, including the privatization of land. In December 1990, the Russian authorities abolished the state monopoly on land and allowed farms to assume control over their assets. At the same time, individual peasants were permitted to leave the state or collective farm on demand with their share of its assets and to establish private farms. Private farms could also be established on unused land, or on land reserved for this purpose, and suitably qualified individuals, including city dwellers, retirees, military veterans and others, could apply to farm there. Restrictions still applied on the sale of agricultural land and there was considerable opposition in the countryside to its full privatization. President Yeltsin has attempted to abolish these restrictions on several occasions, beginning in 1993, but the whole issue continues to be a bone of contention.

Meanwhile, the Russian leadership had decided to reform the collective and state farms. A decree of December 1991 required all collective and state farms to reregister by the end of 1992, and the land and assets of these farms were to become the property of their members. The decree offered the farms four options: (a) they could go into liquidation and sell their assets to the state or someone else (liquidation was compulsory for loss-making farms); (b) they could divide themselves up into private farms; (c) they could sell themselves to some enterprise or organization and thus become a subsidiary farm; (d) they could reorganize themselves as limited liability partnerships, joint stock societies or agricultural producers' cooperatives. The last named forms were most like the collective and state farms and proved the most popular. In March 1992, a fifth possibility was permitted, namely that the collective and state farms could retain their former status, providing they reregistered and assumed ownership of their land.

In general, it can be said that the progress of the reforms has been disappointing to the radical reformers. Most of the land remains under forms of collective ownership, which probably means very little change in the way farming is done, despite the switch in legal status. In fact, both politically and socially the countryside remains very conservative and is usually dominated by farm managers and officials who derive from the old regime (Van Atta, 1994). Years of outmigration by the young and enterprising have left a population which is frequently elderly and conservative, used to the Soviet way of doing things and unwilling or unable to take the risk of establishing private farms. Despite the legal changes, those wishing to secede from the old farms may face hostility from their neighbours or local officials and politicians who naturally oppose developments threatening to their own positions. Given the character of much of the Russian countryside, private farmers may find themselves located miles from the nearest road and far from services and infrastructure. They may also face the possibility of being denied access to services which were traditionally provided by the collective and state farms. Those who establish individual farms often find it in their interests to maintain close relations with the big farms, and a degree of interdependence has often grown up between them. Other factors also militate against the establishment of private farms. These include inflation, low agricultural prices, lack of credit facilities and difficulties in marketing products and obtaining supplies. Private farming has thus failed to transform the Russian landscape, and the rate of founding of new farms has slowed down considerably recently. Such farms are more common in the fertile south than in less fertile regions (figure 7.4; table 7.11).

Table 7.12 shows the contribution made to agricultural output by the three principal types of production unit in Russia: collective forms of enterprise, private farms and personal plots. It will be seen that the output of private farms is still negligible, while the contribution of personal plots is still very important for the more intensive kinds of production. As noted already, the latter is a product of the Soviet era, when peasants were permitted to have plots of land for their own subsistence. Later this private form of agriculture took on significance as a contributor to numerous types of intensive output, and more recently its importance has grown as peasants have been able to acquire more land for this purpose, and as city dwellers and others have been allowed to acquire land to grow food. The recent economic problems have thus encouraged a return to the land. Some commentators argue that this trend to a semi-subsistence form of existence shows that the economic transition in Russia is by no means a transition towards advanced capitalism.

Table 7.13 shows the overall trends in agricultural output since 1981. Agriculture has clearly been unable to escape the problems of the economy in general, and the situation is very disappointing for the reformers. The 1995 grain harvest was the worst since 1963, and 22 per cent below that of 1994. Russia still needs to import food and animal feed, though the demand for the latter has fallen as a result of a severe contraction in livestock numbers (table 7.14). The overall effect has been a deterioration in the Russian diet. Agriculture remains very dependent on state subsidy, though such subsidies are clearly inadequate.

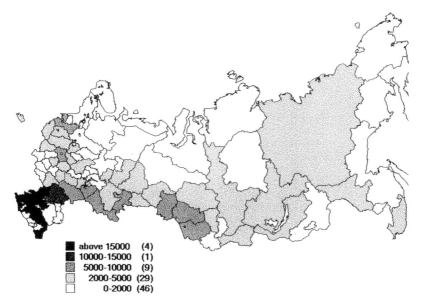

■	above 15000	(4)
▓	10000-15000	(1)
▒	5000-10000	(9)
░	2000-5000	(29)
□	0-2000	(46)

Figure 7.4   Russia: distribution of private farms, 1996.
Source: Goskomstat (1996, pp. 1010–12).

Table 7.11   Private farms, 1991–6

	1991	1992	1993	1994	1995	1996
Number (thousands)	4.4	49.0	182.8	270.0	279.2	280.1
Area (thousand ha)	181	2,068	7,804	11,342	11,870	12,011
Average size (ha)	41	42	43	42	43	43

Source: *Rossiyskiy statisticheskiy yezhegodnik* (1996, p. 555).

Table 7.12   Percentage of total agricultural production by value contributed by types of production unit

	1970	1980	1990	1991	1992	1993	1994	1995	1996
Collective units	69	71	74	69	67	57	54	54	52
Private plots	31	29	26	31	32	40	44	44	46
Private farms	–	–	–	–	1	3	2	2	2

Source: *Rossiyskiy statisticheskiy yezhegodnik* (1996, p. 550), *Russian Economic Trends* (1997, no.1, p. 104).

Table 7.13 Output of key agricultural products, 1981–96

Product	Units	1981–5	1986–90	1991–5	1991	1992	1993	1994	1995	1996
Grain	million tons	92.0	104.3	87.9	89.1	106.9	99.1	81.3	63.4	69.3
Potatoes	million tons	38.4	35.9	36.8	34.3	38.3	37.7	33.8	39.9	38.5
Vegetables	million tons	12.1	11.2	10.2	10.4	10.0	9.8	9.6	11.3	10.7
Meat	thousand tons	8,075	9,671	7,550	9,375	8,260	7,513	6,803	5,796	5,400
Milk	million tons	48.7	54.2	45.4	51.9	47.2	46.5	42.2	39.2	35.7
Eggs	billions	43.1	47.9	40.3	46.9	42.9	40.3	37.5	33.8	31.5
Wool	thousand tons	221	225	151	204	179	158	122	93	

Source: Rossiyskiy statisticheskiy yezhegodnik (1996, p. 572), Russian Economic Trends (1997, no. 1, p. 103).

Table 7.14 Livestock numbers (millions of head),
1970–97 (beginning of year)

Year	Cattle	Cows	Pigs	Sheep and goats
1970	49.4	20.4	27.4	63.4
1980	58.6	22.2	36.4	66.9
1985	60.0	22.0	38.7	64.5
1990	58.8	20.8	40.0	61.3
1991	57.0	20.5	38.3	58.2
1992	54.7	20.6	35.4	55.3
1993	52.2	20.2	31.5	51.4
1994	48.9	19.8	28.6	43.7
1995	43.3	18.4	24.9	34.5
1996	39.7	17.4	22.6	28.0
1997	35.8	16.2	19.5	23.6

Source: *Rossiyskiy statisticheskiy yezhegodnik*
(1996, p. 570), *Russian Economic Trends*
(1997, no. 1, p. 103).

In 1996, three-quarters of farms are said to have run at a loss, and in only three regions (Krasnodar, Stavropol and Bashkortostan) was the farm sector profitable. Marketization is having spatially uneven consequences, as in the cities. Clearly, the establishment of an infrastructure suitable to a market economy, including the physical infrastructure of roads, services and marketing facilities (farms are likely to be increasingly unwilling to provide such infrastructure in the traditional way), will be needed before there can be much hope of improvement. And this will take much investment over a lengthy period.

The changes in the countryside have had differential effects on the rural population. At one end of the scale there is evidence that the rural elite, the officials and managers of the Soviet era, have been able to benefit from policies of privatization, in some cases acquiring large farms and estates as well as interests in the newly privatized services, processing plants and other facilities. Since Soviet servicing and processing plants were commonly developed to serve a handful of large farms occupying huge areas, their acquisition by individuals can often mean the establishment of private monopolies. Given the general lack of infrastructure in the countryside, marketization is likely to lead to the development of regional markets, at least in the first instance. There is some evidence in urbanized regions of food processing industries attracting investment and establishing links with local farms (Hanson, 1996b). Whether such developments will ultimately lead to a genuinely marketized agriculture over considerable areas remains to be seen. If so, rural populations are likely to benefit from the wealth so generated, although the outlook for pensioners and other dependents seems less promising. Elsewhere, the higher costs of fuel and transport associated with marketization have already increased rural isolation, leading to the closure of services and apparently growing poverty (Gur'ianova, 1996). The collapse of collective and state farms, the closure of unprofitable enterprises and the need to pay for services once provided free of charge by the state have increased the misfortunes of many.

Soviet policy produced its own brand of rural misery and inequality, and particularly exacerbated differences with the city. What is uncertain is whether such problems can now at last be addressed, or only grow worse for many of the inhabitants of the Russian countryside.

Eight

The Regions of Russia

For most of its history Russia has been highly centralized. Certain periods, however, stand out as times of decentralization, notably the Time of Troubles in the early seventeenth century, and the years of revolution and civil war beginning in 1917. Such periods occurred when Moscow lost control of peripheral regions, producing anarchy and civil strife. The period since 1990 has also been one of relative decentralization. However, this most recent assertion of regional autonomy has so far taken place without the wholesale descent into disorder which characterized earlier episodes. There are hopes, at least among the democrats, that it may become a permanent feature of the new Russia. Thus it could be argued that there has never been a time when it was more important to know about Russia's regions than today. This chapter will consider the evolution of Russia's regions, and then give more detailed consideration to each major economic region in turn.

Russia's Regions in History

As a general observation it would be true to say that Russia is a country without strong regional traditions (Novikov, 1997). Whereas most western European states arose by welding together pre-existing feudal territories, which often retained their distinctive characters and sometimes certain rights even after their absorption by the unified state, the situation in Russia was quite different. In absorbing its surrounding principalities in the late medieval period, Moscow proved adept at undermining the regional standing of the former princes and boyars (lords) and in tying them to the service of the Muscovite state. No ancient provinces with their special peculiarities, or old aristocracies with the potential to challenge the power of Moscow, were thus tolerated.

Later, when Russian territories expanded beyond the 'homeland', the tsars were careful to ensure that the new regions were settled and administered in ways which would uphold their power. If the new territories were already populated mainly by non-Russians, relatively little effort was made to give them a special status, unless exceptional circumstances demanded it (Raeff, 1971).

When it came to having to devise an administrative framework for their empire, the tsars in effect treated it like an isotropic plain. Down until the time of Peter the Great, the prevailing administrative unit was the *uyezd*, the town and its surrounding district. For reasons of administrative efficiency and internal security, Peter decided to divide the empire into a series of large provinces. However, he did so in a fairly arbitrary way, since there was no pre-existing provincial system to guide him and the network of towns which might act as administrative centres was very sparse (Pallot and Shaw, 1990, pp. 251–6). Later in the eighteenth century, the system of provinces was reorganized by Catherine the Great, but once again some rather arbitrary decisions seem to have been made. With certain modifications, this system lasted down to 1917.

After the 1917 revolution, the Soviets decided to restructure the country's administrative system to reflect their priorities in industrialization and the building of socialism (Pallot and Shaw, 1981, pp. 56–72). Oddly enough, after some changes in policy, the system of oblasts they ended up with bore some resemblance to the old tsarist provinces, at least in European Russia. But detailed changes continued to be made down to the end of the Soviet period, as if to admit that the system had no real historical *raison d'être*. The one major administrative innovation which the Soviets made was in the system of ethnic territories (see chapters 3 and 9), which had no parallel under the tsars.

The point to be made, then, is that both tsarist and Soviet governments felt free to change the administrative units as it suited them and seem to have paid little attention either to historical factors or to local opinion. The Soviets, however, were more careful when it came to tampering with ethnic units, no doubt for fear of stirring up ethnic feeling. As we have seen already, they may thus unwittingly have provided a territorial basis for the gradual development of ethnic and national identity, even where none had existed previously. As far as the non-ethnic units are concerned, the political stability of the long Brezhnev years allowed regional elites to form which established roots in their local areas. In many cases these former Soviet elites are still in power locally. Thus the oblasts as well as the ethnic units inherited from the Soviet era may no longer be as easy to reorganize as they once were, despite calls for their reform from certain quarters (Walker, 1992).

Whether the non-ethnic oblasts are now the object of popular loyalties is difficult to say, but what can be said is that the question of regional autonomy is still a live issue for the Russian Federation, as noted in chapter 3.

Although the Muscovite and Russian imperial states paid little attention to regional differences, this did not mean that the territories over which they ruled were uniform in character. Ethnic variation aside, it may be that the centralization of the state, its continual expansion, its open frontiers and the frequent movement of its peoples (which neither tsarist nor Soviet governments found easy to control) helped to smooth regional variations. Even so, the natural environment, historical circumstances and way of life of the people have not allowed them to be entirely eradicated. The northern part of European Russia, for example, has long been noted for its distinctive character in association with its environment (predominantly boreal forest and tundra) and its mainly non-agricultural economy. Historically, serfdom never spread into the north, and that fact, together with its trading links with the outside world via ports like Arkhangel'sk on the White Sea, made it different from (and possibly freer than) much of Russia. It was the European north which was the springboard for the colonization of Siberia, and some of the region's characteristics (like the absence of serfdom) transferred themselves there. Siberia's sense of separateness from European Russia (the latter generally referred to as Rus') was reinforced by the fact that the tsars long kept it apart as a fur colony and then as a penal colony. The Soviets regarded it in a similar way – as a provider of resources and also as a place for the exile of undesirables. The Siberian mentality is still defined by a sense of being on the frontier, as well as of having been mistreated or ignored by the rulers in far away Moscow. The European south, by contrast, has been defined by its formerly agricultural character (the black earth regions were and are Russia's breadbasket and the regions where serfdom reached its greatest development). It also had an historical role as the defensive frontier against the tsar's enemies. Cossack traditions, which have recently been revived in some areas, sometimes express themselves in opposition to Moscow's influence, and sometimes in energetic and even violent Russian nationalism.

The democratization of Soviet and now Russian politics has allowed scholars to glimpse differences in values and outlook across Russia in ways which were impossible in the past. For example, since the 1989 Soviet elections to the Congress of People's Deputies, electoral geography has attracted the attention of numerous scholars of Russia, providing important clues to regional differences in culture and outlook. Of course, election results are influenced by a multitude of factors, many of which change through time, and it would be quite unrealistic to expect them to

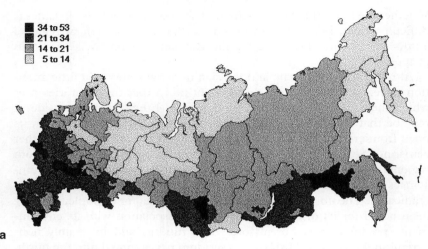

■	34 to 53
▓	21 to 34
▒	14 to 21
□	5 to 14

a

Figure 8.1　Russian parliamentary election, 1995: vote for (a) communist and (b) reform parties.
Source: after Clem and Craumer (1995, pp. 597–600).

reflect differences in regional culture or outlook in any straightforward way. Nevertheless, the studies which have been conducted since 1989 do point to the persistence of broad regional patterns in voting behaviour in the several elections that have been held since then (Clem and Craumer, 1995, 1996). Support for the parties of reform (parties which support President Yeltsin's policies or which support rather similar programmes if opposed to him) tends to be strongest in regions across northern Russia, particularly in the cities of Moscow and St Petersburg, but also in parts of the North and Northwest, some areas in the Central Region, Nizhniy Novgorod Oblast in the Volga–Vyatka Region, some ethnic units along the Volga, some of the industrial oblasts in the Urals, across ethnic units in northern Siberia, and in the Far East. By contrast, the left-leaning parties (communists and others) tend to find support in the southern tier of regions, from Smolensk in the west via the Central Black Earth Region, parts of the North Caucasus and the southern part of the Volga Region, to continue across southern Siberia and the Far East (figure 8.1). Support for the parties of the right is more scattered across the regions.

Looking at the results of the December 1995 election and earlier returns, Clem and Craumer found that support for reform parties tended

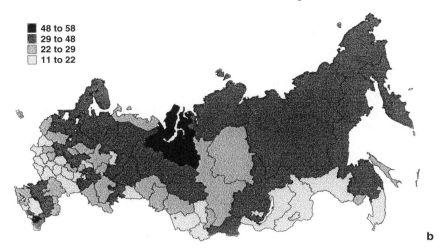

48 to 58
29 to 48
22 to 29
11 to 22

b

Figure 8.1 *Continued*

to be positively correlated with higher levels of urbanization, education and percentages of the workforce engaged in 'mental' work, and negatively with increasing numbers of agricultural workers and old people. The reverse applied to the left-leaning parties, while support for the parties of the right displayed no clear correlations, though it did tend to rise with higher percentages of the population consisting of ethnic Russians. Another study, by O'Loughlin et al., underlines the significance of ethnicity in explaining regional differences in parliamentary election results. The authors note the overriding importance of centre–periphery relations and see a growing role being played by the historic divide between European and Asiatic Russia (O'Loughlin et al., 1996).

From the eighteenth century, geographers and others busied themselves devising schemes of large-scale regionalization of Russia, taking into consideration such factors as the natural environment, population, economy and settlement. Often these endeavours had a practical basis, arising out of the desire to make sense of this vast country and to have an inventory of its resources and potentials for development. Sometimes, it was argued that the schemes should form the basis for future political or administrative subdivisions. One of the most famous of these projects was the multi-volume work *Rossiya*, edited by the eminent geographer V. P. Semenov-Tyan-Shanskiy (Semenov-Tyan-Shanskiy, 1899–1914). In principle, this divided Russia into some 22 regions and popularized the idea of such regions as the Central Industrial, Central Black Earth, Volga and Urals. After 1917, the Soviets devised their own schemes of economic regionalization which they hoped to use for planning purposes. Eventually, they divided the Russian Federation into eleven economic regions (see figure 1.3), some of which resembled regions proposed by Semenov and others. This system does provide the basis for popular notions of the major regions of Russia today. However, where oblasts and other units have grouped themselves into regional associations in order to promote their interests in Moscow and other contexts (see, for example, Radvanyi, 1992), they have occasionally overlapped the boundaries of the economic regions. This shows that the economic regional boundaries are not deemed satisfactory for all purposes, and competition between oblasts tends to reduce their potential role as a framework for regional cooperation.

Even so, the economic regions are still used for statistical and other purposes today, and have the benefit of convenience. They will thus form the basis of the discussion which follows, describing each economic region today, the changes they have undergone in the recent past and the problems they face in the future. Any conclusions drawn about a region's potential must be regarded as tentative, however. At the current stage of economic transition it is especially difficult to be certain how the various parts of Russia are likely to fare in the future.

Moscow and the Central Economic Region

Lying at the heart of the historic Russian state (though with a western border which since 1991 has coincided with that of the state), the Central Economic Region (figure 8.2), often referred to as the Central Industrial Region, is the core of the Russian economy. With over 30 million inhabitants, or more than 20 per cent of the total population of Russia, the region accounted for about 16.8 per cent of Russian industrial production in 1995 in an area less than 3 per cent of the total Russian territory. Of the 16.8 per cent industrial production, the city of Moscow accounted for 5.7 per cent and its surrounding oblast a further 3.1 per cent. The entire region is thus focused on Moscow, whose role as the country's capital has been and continues to be a formative influence in its economic structure.

With almost 83 per cent of its population living in towns and cities (compared with a national average of 73 per cent), the region is highly

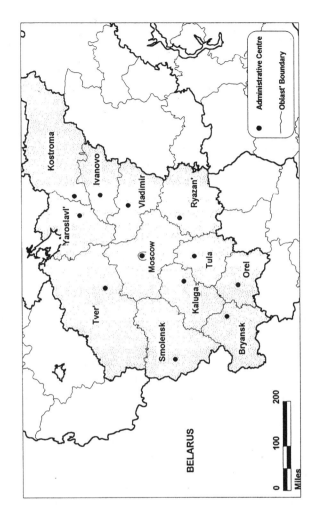

Figure 8.2 The Central economic region.

urbanized. Not all parts of the region are equally developed in this respect, however, and parts of the south (where the region borders on the fertile Central Black Earth Region), the west, northwest and extreme northeast contain areas with relatively few towns and cities. By contrast, to the east and south of Moscow lie some of the oldest industrial regions in Russia. Located in the mixed forest vegetation zone, the Central Region's soils are not particularly fertile, with the exception of parts of the south. The rather complex topography and the effects of glacial action have given rise to a varied rural landscape, with forests, marshes, lakes and agricultural land. The region is not noted for its mineral resources and the Moscow brown coal basin yields only low grade coal.

The Central Region's industrial role dates back as far as the seventeenth century, when various proto-industries began to arise to serve the needs of the growing capital and the state. By the next century, the region had begun to develop a textile specialization, but many small-scale crafts and activites had also begun to arise to serve the slowly expanding market. With the onset of Russia's industrial revolution in the latter half of the nineteenth century, these activities continued to expand and others, including engineering and associated branches, began to join them. The Soviets naturally regarded the region as pivotal to their industrialization policies, and it became noted as a region with a wide variety of industrial activities, including machine building, chemicals, food and light industries. The central role played by the region in the Soviet economy was reflected in the importance given to the military–industrial complex. The region was also of great significance to the Soviet Union's drive for technological supremacy, and a considerable proportion of the country's leading scientific research and design institutes were located there. About a third of all the country's workers in science and science-related activities lived in the Central Region, particularly in the city of Moscow and Moscow Oblast.

The very fact that the Central Region was so closely identified with the industrial priorities of the Soviet era means that it will be poorly placed as a region to adjust to post-Soviet conditions. The development of its industrial structure since 1991 reflects the economic changes which have occurred since the fall of the USSR and which were discussed in chapters 4 and 5. Thus many parts of the machine building industry are now suffering as a result of structural change, and especially from the effects of reduced spending on defence. The textile industries which have been of particular importance to the region have also suffered eclipse. Reduced domestic demand, competition from cheap imports and supply difficulties appear to be the major factors here. Another problem which affects most industries in the region is rising prices for fuel and raw materials, the effects of which are likely to be especially severe in an area without abundant natural resources of its own. Up to 80 per cent of the region's light industry enterprises are potentially bankrupt. The industrial regions to the east of Moscow seem to be most seriously affected by various indicators of industrial crisis and decline, especially Ivanovo Oblast, which is a centre for the textile industry, but also the eastern parts of Moscow Oblast, Vladimir, Yaroslavl', Kostroma and also Bryansk to the southwest of the capital. Where agriculture is rather more important (e.g. Smolensk), where there is a degree of mining and raw materials processing (e.g. Tula) and where there is the advantage of proximity to Moscow (parts of Moscow Oblast), the situation is slightly better. But local factors – for example, whether or not there are state orders for defence industries or the policies being pursued by local government – also have some importance.

The city of Moscow and its immediately surrounding region present a contrast with this generally depressing picture. In chapter 5, the city was classified as a 'hub' region. This derives from its position at the centre of a web of national and international routes and from the presence in the city of the organs of central government, always an important factor in the history of Russian economic development. With 8.8 million people in the city plus another 6.6 million in the oblast, the entire capital region is a major market containing just over half of the Central Region's total population. As noted already, the city and oblast also make a significant contribution to the region's industrial output.

As former linch pin of the Soviet economy, Moscow has been unable to escape the effects of restructuring. Thus industrial output in 1994 stood at 46.8 per cent of that in 1991. Traditionally, the city's industries included a wide range of machine building branches (including automobiles and aircraft), metalworking, chemicals, light industries (especially textiles) and food processing. In keeping with the inherent bias of the Soviet economy, heavy industry played a bigger role than would be expected in Western metropolitan centres of comparative size and importance. Many of these industries, often forming part of the military–industrial complex, have been suffering in the recent period, as have the light industries. To offset this picture, however, the city has also witnessed the rise of many small enterprises and joint ventures whose output tends to escape the official statistics, and therefore the fall in industrial production is likely to have been far less than the official figures suggest. What has been particularly remarkable about Moscow's case has been the economic restructuring which has taken place (Gritsai, 1997). Thus, whereas statistics suggest that between 1990 and 1993 the city lost 7 per cent of its industrial em-

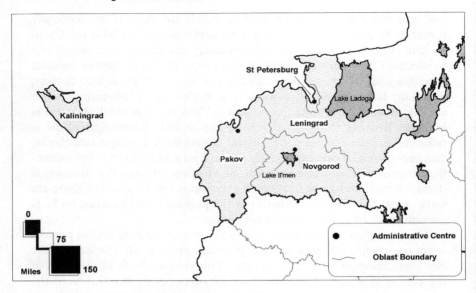

Figure 8.3   The Northwest economic region.

ployment, 19 per cent of that in transport and communications and 11 per cent in education, culture and science, there were major gains in the financial sector (129 per cent), administration and government (30 per cent) and wholesale and retail trade (30 per cent). As noted in chapter 5, the city has become Russia's principal centre for financial and business services, and has a major role to play in attracting foreign investment and establishing joint ventures, especially in the retail, consumer and service sectors. The burgeoning economy is reflected in the relatively high average income levels, the booming construction industry, dramatic rises in car ownership levels and the growth of crime. At the same time, the city's antiquated infrastructure can barely cope with the rapid changes and there is a pressing need for investment in ground and air transportation, telecommunications and other facilities if growth is to continue.

St Petersburg and the Northwest Region

The Northwest Region (figure 8.3) consists of the city of St Petersburg (Leningrad during most of the Soviet era), and the three oblasts of Leningrad, Novgorod and Pskov. The region has a population of 8.1 million, 60 per cent of whom live in St Petersburg and a further 21 per cent in Leningrad Oblast. In 1995, with 5.5 per cent of the country's population, the region accounted for 4.2 per cent of Russian industrial production, with St Petersburg's share standing at 2.5 per cent.

Most of the Northwest Region forms part of the original Russian homeland, although the area bordering on the Gulf of Finland (known as Ingria) was long disputed and was finally annexed to Russia by Peter the Great at the beginning of the eighteenth century. This was where Peter decided to build his new capital city of St Petersburg, which was designed to imitate the European cities the tsar so much admired and meant to expose Russia to European ideas and technology. St Petersburg remained capital until 1918, and the city's entire outlook and orientation continued to form a pointed contrast with those of Moscow. Even today there remains a rivalry between the two cities, St Petersburg with its northwestern and seabord location seeming to offer a more European or westward-looking destiny for Russia than continental Moscow.

As capital of the tsarist state, St Petersburg's growth was rapid, and in the nineteenth century the city participated fully in Russia's industrial revolution. By 1917, St Petersburg had become a major centre for machine building and light industries, with many port-related and government-oriented activities. After the Russian Revolution, with the transfer of the capital to Moscow, the city was relatively neglected by the Soviet government, and Stalin in particular seems to have nursed a grievance against the city and its leaders. Moreover, the inward-orientation of the command economy under Stalin and later did much to undermine the city's historic role as one of the country's leading ports, while much damage was done during the Second World War. Even so, the Soviets developed the major industries of the city for their own ends, with a focus on machine building, shipbuilding and repair, light industries and the manufacture of consumer goods. A particular feature of St Petersburg and the surrounding region has been the importance of the military–industrial complex. It is estimated that, in 1989, 34.4 per cent of industrial employment and 27.8 per cent of industrial output in the city were oriented to the military–industrial complex. In Novgorod Oblast, whose industries, like those of Leningrad Oblast, are closely tied to those of St Petersburg, the corresponding figures were 39.2 and 45.4 per cent respectively (Bradshaw, 1996, p. 24). Other activities in the region include aluminium smelting at Volkhov in eastern Leningrad Oblast, forestry and agriculture. As a borderland between the boreal forests and the mixed forest zone, the region's soils are not very fertile and much agricultural activity has focused on livestock (which has suffered recently) and urban-oriented production. Other than timber, the region is poorly endowed with natural resources.

Given its industrial profile and marked dependence on the military–industrial sector, it is inevitable that the Northwest will suffer as a result of restructuring. Thus total industrial output in 1994, according to official figures, was only 51.2 per cent of that in 1991, while light industries had declined to less than 30 per cent. However, as noted in chapter 5, St Petersburg in particular, and by implication some surrounding centres, constitute a 'gateway' region with numerous potential advantages for future development. The latter include the presence of a large market, a skilled labour force, an important tradition in science and technology, and proximity to the western frontier and sea coast. The loss of ports in the Baltic states is likely to enhance St Petersburg's importance as gateway to the outside world, and about 30 per cent of Russia's imports and 25 per cent of its exports pass through the city. However, the port is badly managed and is in great need of investment, while border crossings into Finland likewise need upgrading. High speed rail links with Helsinki and Moscow are under consideration, and the city's airports at Pulkovo are currently being expanded and modernized. Although many of St Petersburg's industries are old and failing, the city with its history and European ambience is proving an attractive location for Western investment and the establishment of joint ventures. Firms unwilling to pay the high prices now prevailing in Moscow seem to be finding St Petersburg a suitable alternative. Foreign interest embraces such sectors as food processing, tobacco, pharmaceuticals and tourism. The city also has a growing financial and business services sector, though well behind that of Moscow.

The economic future of the Northwest Region is obviously closely bound up with that of St Petersburg, but areas closer to the city seem most likely to feel the positive benefits first. Chapter 5 noted how Pskov Oblast, away on the Estonian border, has felt the negative effects of restructuring most severely. This is a mainly agricultural region.

A region which might usefully be included within the Northwest is Kaliningrad Oblast, a Baltic exclave now cut off from the rest of Russia by Lithuanian territory. The region is the northern half of what was once German East Prussia, which was annexed by the USSR in 1945 and had its German population expelled. The region is attempting to combat its isolation by developing its import and export functions for Russia, its food processing industries and other activities. It became a free economic zone in 1991, but a large military presence may militate against foreign involvement.

The Northern Economic Region

Soviet policy generally favoured the linking of northern territories with more developed southern ones, and the Northern Region (figure 8.4)

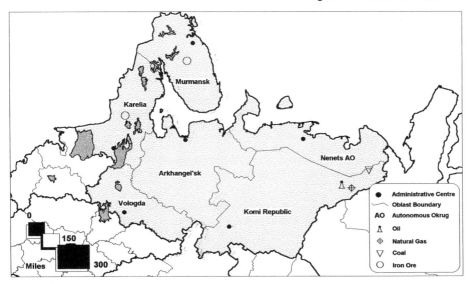

Figure 8.4   The Northern economic region.

was only separated from the Northwest in 1982. It is a huge area, occupying 8.6 per cent of the national territory, though containing only 4 per cent of the population. The region consists of two republics of the federation (Karelia and Komi) and four oblasts. Arkhangel'sk Oblast contains the Nenets Autonomous Okrug. In the 1989 census, 87 per cent of the population of the Northern Region were ethnic Russians. The region is 76 per cent urbanized, compared with a national level of 73 per cent.

The Northern Region's economy reflects two Soviet priorities. One was the development of natural resources for reasons of domestic industrialization and, somewhat later, for export. In 1993, the region accounted for 7.8 per cent of the output of Russia's extractive industries, compared with only 5 per cent for industry as a whole. Extractive industries include forestry, non-ferrous metals, ferrous metals and fuels. The processing of raw materials is important in numerous locations, while the southern oblast of Vologda, being closer to Russian markets, has an economic profile which includes iron and steel, engineering and chemicals. Fishing and fish processing are important in Murmansk, Arkhangel'sk and other ports. However, since the Northern Region largely coincides with the boreal forest and the tundra, agriculture plays a relatively minor role. The other Soviet priority was defence and military activities, in association with the region's proximity to the northern seas and the USSR's western border. The Kola peninsula was a highly militarized region and the various northern ports developed numerous supporting functions.

In the new conditions now prevailing, the Northern economy has been suffering from the effects of rising costs of transport and energy, and reduced investment for defence-related activities. Some high cost mining activities are now having to close, while considerable investment is needed for improvements in transportation and cleaning up the environment. Pollution produced by some northern processing plants has had an international impact, and the Scandinavians and others have been keen to find solutions. Areas focused on the military are having to revise their economic profiles. The Northern Region's accessibility to northern and western Europe and its rich endowment of resources have proved beneficial for the export of natural resources and for the attraction of foreign investment. The number of joint ventures is growing, while foreign interest or involvement has targeted onshore and offshore energy production, forestry, transport and telecommunications. There is considerable interest in the possibility of oil and gas production from the Barents Sea shelf. In conclusion, while this region seems to have potential as a raw material exporter and processor, much investment is needed to exploit that potential and to redirect its economy away from the priorities of the past.

The Volga–Vyatka Region

The Volga–Vyatka Region (figure 8.5), which lies to the northeast and east of the Central Region, is very much a transitional zone between the boreal forests in the north and the forest-steppe in the south, and its economic activities reflect the influence of its neighbouring economic regions. Thus Kirov Oblast in the north shares some of the characteristics of the Northern economic region, the Mordvinian Republic in the south has soils and natural vegetation similar to those of the Central Black Earth Region, to which it was attached until 1960, whereas Nizhniy Novgorod Oblast can be regarded as an eastern extension of the Central Region. Topographically it is a region of mixed relief dominated by the Volga and its tributaries.

Culturally, the Volga–Vyatka Region is also very much a transitional zone, forming an historical meeting point between Finno-Ugrians (Mari, Mordvinians) and Russians on the one hand, all of whom were originally forest-dwelling peoples, and the Hunnic Chuvash who penetrated the region from the steppe. Part of the area falls within the historic Russian homeland (notably Nizhniy Novgorod) and the rest was colonized by the

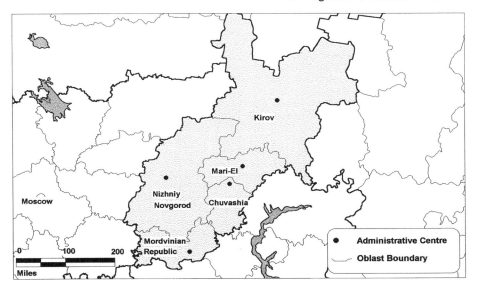

Figure 8.5   The Volga–Vyatka economic region.

Russians in the course of their penetration of the boreal forests and the forest-steppe from medieval times onwards. The ethnic mixture is today reflected in the political geography of the territory, which consists of two oblasts (Nizhniy Novgorod and Kirov) and three republics of the Federation (Mari-El, Chuvashia and the Mordvinian Republic). Russians constitute the overwhelming majority of Volga–Vyatka Region's population (75 per cent), but they fall to less than half in the case of Mari-El and to almost a quarter in that of Chuvashia.

Volga–Vyatka is one of the smallest of Russia's economic regions, with only 1.6 per cent of the country's surface area but 5.7 per cent of its population (8.5 million people). In terms of both population and economy it is very much dominated by Nizhniy Novgorod Oblast, which contains 44 per cent of the region's total population of nearly 8.5 million people. The city of Nizhniy Novgorod (formerly Gor'kiy), with 1.4 million people, is the third biggest city in Russia. The economy of Volga–Vyatka reflects the region's transitional character and also its strategic location as a crossing point for east–west routes. Hence it has benefited from its accessibility to the fuels of the east and to the markets of the west, its own lack of resources thus being largely counteracted. Among its traditional activities are engineering (based on Nizhniy Novgorod

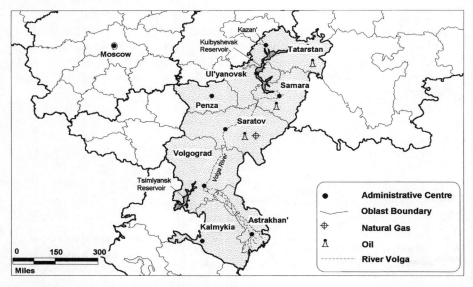

Figure 8.6   The Volga economic region.

with its major Gor'kiy Automobile Plant, but many other activities spreading into the republican capitals), woodworking (especially in Kirov), chemicals and petrochemicals, and food and light industries. Agriculture's importance varies with natural conditions, but it suffers from low investment and other problems, and the rural areas have traditionally suffered from outmigration.

The significance of engineering, having close connections with the military–industrial complex, bodes ill for the region's economic future. Several of Volga–Vyatka's subdivisions, especially the three republics, have been suffering severely in the transition. However, the region's strategic location and the dynamic leadership of some local politicians (notably Boris Nemtsov, former governor of Nizhniy Novgorod) have attracted some outside interest, and the city of Nizhniy Novgorod may yet spearhead a more buoyant economy in the future.

The Volga Region

The Volga Economic Region (figure 8.6) lies astride the north–south flowing Volga River, historically one of Russia's major highways. It is in fact the river which gives the region its unity, since there is rather little basis for unity in other respects. In terms of physical geography, for example, the region can be divided into four or five geomorphic zones, and the soils and vegetation change gradually from the mixed forests of part of Tatarstan in the north through the forest-steppe and steppe to semi-desert and even pure desert around the Volga delta. There are also considerable economic differences between the more industrialized and urbanized north and the more agricultural south. Altogether, the Volga Region is home to 16.9 million people (11.3 per cent of Russia's population) and in area is the sixth largest economic region in the country.

Russia's occupation of the Volga River and its environs followed Ivan the Terrible's seizure of the Tatar capital of Kazan' in 1552. The area was subsequently mainly settled by the Russians, thus becoming part of the 'settlement empire' (see chapter 1), but some important elements of the non-Russian population have remained. In terms of the political geography, the region embraces two republics: Tatarstan (with about 48 per cent of its population Tatar and 45.5 per cent Russian) and Kalmykia-Khal'mg-Tangch (with just over half of its population believed to be Kalmyks). There are additionally six oblasts with predominantly Russian populations.

Economically, the Volga Region has many advantages. These include a large and predominantly Russian population (making for stability, although some difficulties have been experienced over Tatarstan's independent stance), a strategic location between the industries of European Russia and the resources of the Urals and the east, the Volga River (hydro-electric power, fishing, transportation), good natural resources (agricultural, the Volga–Urals oil and gasfield, natural gas near Astrakhan) and a long tradition of crafts and engineering. It is therefore hardly surprising that the region is fourth among Russia's economic regions for industrial output and second for agricultural production. Its initial economic development took place in tsarist and early Soviet times, but it was the industrial evacuations associated with the Second World War which precipitated its modern importance. Later, in the 1950s and 1960s, came the development of the Volga–Urals field, the damming of the Volga for hydro-electric power and the spread across the region of oil and gas pipelines. The major industrial sectors today are the fuels-energy sector (14.6 per cent of output in 1993), machine building (33.5 per cent), chemicals and petrochemicals (14.2 per cent) and the food industry (10.8 per cent) (Bradshaw, 1996, p. 59). Two of the major industrial developments of the 1960s and 1970s are associated with the motor industry: the Volga Automobile Plant (AvtoVAZ) at Togliatti (accounting for two-thirds of Russian passenger car production) and the Kama Automobile Plant (KamAZ) at Naberezhnyye Chelny in Tatarstan (heavy trucks and passenger cars). Also important is the Ul'yanovsk Automobile Plant for jeep and bus production, which

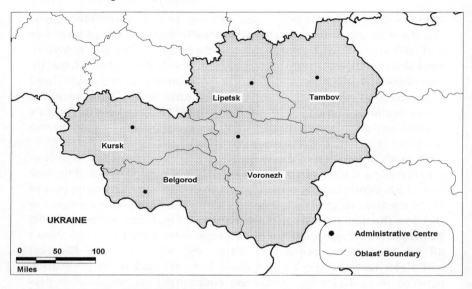

Figure 8.7   The Central Black Earth economic region.

started life as a plant evacuated from war-torn Moscow in 1941. The region's development has been achieved at a cost, however – the pollution of the Volga and the deterioration of agricultural resources in Kalmykia and along the lower Volga were two of the themes highlighted in chapter 6.

While the Volga Region's aerospace and automotive industries may prove attractive to outside investment (and that is by no means guaranteed), other activities in the region are finding it difficult to acquire the capital needed for restructuring. There is a strong local lobby for protectionism of established manufacturing industries, many having a military orientation. Resource-based activities, like Tatarstan's oil industry, may find a more secure future, though the oil and gas wells are near exhaustion in some cases. Economic contraction seems on the cards for many industries, but the region's numerous innate advantages cannot be ignored and may prove significant again in the future.

The Central Black Earth Region

The Central Black Earth Region (figure 8.7) which lies due south of the Central Region in the fertile forest-steppe and steppe vegetation zones, was once Russia's breadbasket. The region was absorbed by the expanding Russian state in the sixteenth and seventeenth centuries, as the tide of Russian defensive towns, military lines and settlement gradually pushed the nomadic Tatars southwards. The population is thus ethnically overwhelmingly Russian. In the nineteenth century, continued population growth reduced many areas to a state of acute rural crisis, and outmigration has been a characteristic feature ever since. With an urban population of just under 62 per cent in 1996, compared with a national average of 73 per cent, agriculture continues to play an important economic role, though still suffering from the effects of Soviet policy. In 1994, the Central Black Earth Region occupied second place among the regions of Russia for per capita agricultural production, and second place for the productivity of grain agriculture. In view of its relatively small population (7.9 million in 1995), it was only seventh for total agricultural production. The rural and agricultural bias do much to explain the region's conservative political stance – it forms part of the much publicized 'red belt' of communist- and nationalist-supporting regions, generally opposed to the economic reform policies of the Yeltsin government.

Having a total area of 167.7 thousand square kilometres, or less than 1 per cent of the national territory, the Central Black Earth Region is the smallest of Russia's eleven economic regions. As its title suggests, it was long considered to be part of central Russia, having a strategic location between Moscow and the industrial regions of southern Ukraine. Since Ukrainian independence, however, this advantage has waned, as the new international border has made movement to the south more difficult and as the border itself has given rise to tensions associated with illegal migration, smuggling and other activities. The region thus threatens to become peripheral to the mainstream of life in the new Russia, with its generally poor infrastructure impeding access to many rural dustricts.

Apart from agriculture and the associated food processing industries, the major economic development of the Soviet era was that of the Kursk Magnetic Anomaly, the enormous iron ore field in the western part of the region which was developed in the 1950s, 1960s and 1970s. This makes a major contribution to Russian iron ore production, and the associated ferrous metallurgical industries are also important. Since the industries are relatively modern, they have attracted a good deal of Western interest. Other activities in the region include engineering and chemicals. These are particularly well represented in the administrative centres. Limited availability of energy and water is a problem for industrial expansion.

In conclusion, the Central Black Earth Region appears to be an area

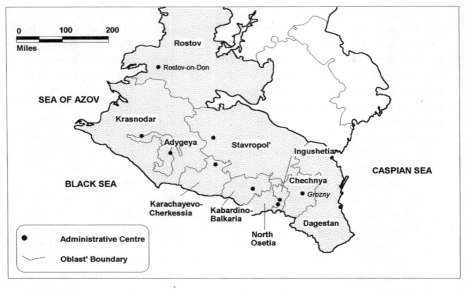

Figure 8.8   The North Caucasus economic region.

of mixed fortunes, with the metallurgical industries performing relatively well on the international market but other activities suffering problems. Local living standards have to some degree been protected by lower than average food prices, no doubt helped by the cultivation of private plots.

The North Caucasus Economic Region

Since the break-up of the USSR in 1991, the North Caucasus Region (figure 8.8) has formed a strategic segment of Russian territory, running between Ukraine, the Sea of Azov and the Black Sea to the west, the lower part of the Volga region (and just beyond that, Kazakhstan) and the Caspian Sea to the east and down to the crest of the Caucasus and the Transcaucasian states to the southeast. Historically, this area has been a crossing point of routes, and it remains so to this day. It is now particularly significant as providing Russia's only access to the Black Sea and hence the Mediterranean basin.

Much of the region was occupied in the past by nomads and cossacks who roamed the steppe grasslands. Russian settlement penetrated the region from the sixteenth century, but had the greatest impact from the eighteenth. The Russians had particular difficulty controlling the Caucasian mountain peoples, however. The area long remained one of ethnic confrontation, and since 1991 has become so again, particularly in the case of the Chechen war. The ethnic complexity of the mountainous territory and its adjacent lowlands is reflected in the political geography. The economic region consists of no fewer than seven republics, located mainly along the Russian side of the Caucasus range and embracing some of the lowlands to the north, the two krays of Krasnodar and Stavropol', and Rostov Oblast in the northwest. The lowlands are mainly settled by the Russians (originally by many Ukrainians as well), and thus the non-ethnic units are mainly Russian in population. These three units account for over two-thirds of the territory and the population of the region. Of the republics, which vary in size and population, only Adygeya and Karachayevo-Cherkessia had majority Russian populations in 1989. Altogether, in 1995 the North Caucasus had a total population of 17.7 million and occupied 2.1 per cent of the national territory.

Climatically, the North Caucasus has one of the most favourable climates in Russia for agriculture; only towards the east does the climate become drier, particularly northeastwards in the direction of the lower Volga. The region is thus one of Russia's major agricultural producers, coming fourth among economic regions in value of agricultural production (1994). In 1993, it was the first grain producer, with the highest grain yields in Russia. Krasnodar and Rostov are particularly important agricultural producers. Food and light industries are significant activities in this region. Others include engineering, oil refining, chemicals, wood processing and, in the past at least, tourism. The Donbass coalfield lies partly in Rostov Oblast, a relatively small but strategic producer.

The North Caucasus has traditionally been one of Russia's least economically developed regions, and it has suffered acutely from the political instability and conflicts of the post-1991 period. An influx of refugees and other migrants has added to its problems. The republics have been experiencing particular economic difficulties and, as noted in chapter 5, are among the most distressed regions in Russia. Some of the region's severe environmental difficulties have been commented on in chapter 6. Things seem rather brighter in the western parts of the region, where foreign investment has been attracted into food processing, telecommunications and other activities. A more stable political climate and environmental improvement will considerably enhance prospects for regional development. In the longer term, this region can hope to benefit from its climatic advantages, its proximity to the Black Sea and the possibilities which exist for commercial and other linkages with neighbouring countries. The coastal regions look set to become Russia's 'gateway' to the Mediterranean and beyond.

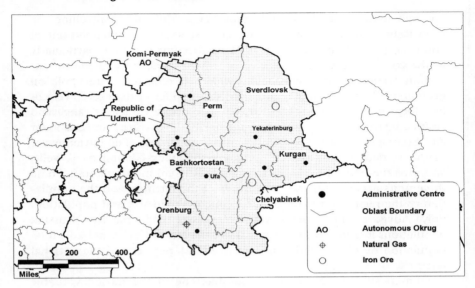

Figure 8.9   The Urals economic region.

The Urals Economic Region

In sharp contrast to the underdeveloped North Caucasus Region, the Urals Economic Region (figure 8.9) was one of the powerhouses of the Soviet economy. The region embraces the central parts of the Urals range, but not its most northerly parts, which form a borderland between the Northern and West Siberian regions, or its southern extensions, which lie in Kazakhstan. It also includes important territories to the west and east of the range. The Urals form the borderland between European Russia and Siberia (the mountains themselves are no barrier to movement), a communications link and a rich reservoir of resources. Natural resources, however, are no longer as abundant as they once were. In physical terms, the region runs north–south through several zones of vegetation and soils, from boreal forest to steppe, but there are considerable local variations in association with relief and climate.

While Russians were active in the northern Urals from late medieval times, the region fell under Russian control in the latter half of the sixteenth century, when it was valued for its furs, salt, potash and other resources. In the early eighteenth century under Peter the Great, an iron industry arose, utilizing local ores and timber, but in the nineteenth century the region was outshone by the newer industrial centres of

Ukraine and European Russia. Only with Stalinist industrialization did the region once again blossom as a centre for mining, metallurgy and heavy engineering. Chemicals and petrochemicals were later additions.

In moving into the Urals region, the Russians encountered a number of Finno-Ugrian and Turkic peoples whose presence is reflected in political geography. Today the region consists of two republics – Bashkortostan and Udmurtia – and five oblasts: Kurgan, Orenburg, Perm, Sverdlovsk and Chelyabinsk. The Komi-Permyak Autonomous Okrug lies within Perm Oblast. Only in the latter region are Russians outnumbered by the titular population.

The Urals economic region occupies about 4.8 per cent of the national territory and contained 13.8 per cent of the population in 1995 (20.5 million people). With over 74 per cent of its population living in towns and cities, the region's urbanization level exceeds the national average, but the Urals contains both more highly industrialized (Sverdlovsk, Chelyabinsk, Perm) and more agrarian (Kurgan, Orenburg) regions. The industrial complex of the central and southern Urals is characterized by its dependence on mining, ferrous and non-ferrous metallurgy and heavy engineering. As noted in chapter 5, the region is greatly oriented towards the military–industrial complex, reaching very high levels indeed in the Udmurt Republic. The development of the Volga–Urals oil and gas field from the 1930s led to the rise of chemical and petrochemical industries in Perm and Orenburg oblasts and in Bashkortostan. Other activities in this region include forestry and a growing financial and service sector, the latter centred on the major city of Yekaterinburg (formerly Sverdlovsk). In 1995, the Urals ranked first in order of industrial production among Russian regions, narrowly beating the Central Region.

Given the close link between the economic development of the Urals and that of the USSR, many of the region's industries have inevitably been suffering during the transition period. It is particularly the heavy engineering sectors, so often tied to the military–industrial complex, where this is true. However, other sectors have been performing better than might have been expected, successfully seeking export markets to replace those lost at home. Thus metallurgy, chemicals, petrochemicals and forestry industries have been important contributors to exports. There is also some evidence that defence plants are trying to increase their exports of armaments. Some foreign investment has been attracted into the region. Politically, the region displays some signs of local identity and activism: witness the prominence on the national stage of such politicians as Eduard Rossel, governor of Sverdlovsk Oblast. Thus, while the Urals might appear a good

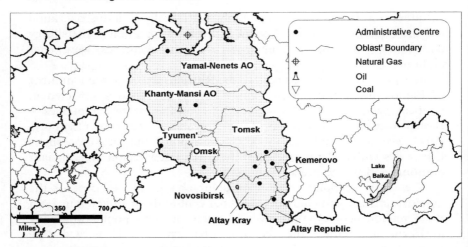

Figure 8.10   West Siberia economic region.

candidate for designation as part of the 'frostbelt', its varied economy and strategic location might mean that it is too early to consign it to history's dustbin.

The West Siberian Economic Region

The three economic regions lying east of the Urals – West Siberia, East Siberia and the Far East – account for three-quarters of the territory of the Russian Federation, but only about 21 per cent of its population (32 million people, nearly half of whom live in West Siberia). In general terms, human settlement diminishes as one moves towards the east, but is particularly low to the north. This has long been European Russia's resource appendage, cut off from the 'motherland' by distance and by a frontier mentality. Nowadays, however, West Siberia (figure 8.10), and especially its southern part, forms part of the European 'ecumene'. In Russian parlance, the term 'Siberia' refers to West and East Siberia, but not to the Far East, thus differing from the way the term is popularly used abroad.

West Siberia's annexation to the Russian state began in the late sixteenth century with the overthrow of the Tatar khanate of Sibir' and the opening up of the northern, largely forested part of the region to the fur trade. Later Russian settlement began to penetrate southwards on to the forest-steppe and steppe at the expense of their nomadic inhabitants. In the late nineteenth and early twentieth centuries, after the completion of the Trans-Siberian Railway, the southern area was subjected to intensive agricultural colonization, a process which continued in the 1950s in the Virgin Lands scheme. Two major industrial projects marked the Soviet era: the development of the metallurgical and heavy engineering industries of the Kuzbass coal basin in the southeast during the 1930s, and some thirty years later the opening up of the oil and gas fields of the north (see chapter 5). Southern cities like Omsk and Novosibirsk, oriented along the Trans-Siberian Railway in the belt of denser settlement, also hosted industries like machine building, while Omsk and Tomsk developed petrochemicals. By contrast, Altay Kray in the southeast is an important agricultural region.

The history of Russian settlement in West Siberia is reflected in its present-day administrative structure. Thus, in the energy-rich north, where Russian settlement is largely confined to the resource centres, lie the two autonomous okrugs of Yamalo-Nenets and Khanty-Mansiy. Both of these are officially part of Tyumen' Oblast, but they have recently been demanding their independence. The titular populations are nowadays only small minorities in both autonomous okrugs, although the Russians have recently been leaving in considerable numbers for economic reasons. Russians also outnumber the titular group in the Republic of Altay in the mountainous southeast, the only republic in West Siberia. All other administrative units are overwhelmingly populated by Russians.

Surveying the West Siberian scene in the mid-1980s, Leslie Dienes argued that, despite its higher level of development compared with other eastern territories of the USSR, the region was suffering from the distorting effects of Soviet-type policies (Dienes, 1987). Thus the machine building industry was underdeveloped and insufficiently geared to local needs, leading to the need to import many necessities from elsewhere, while high levels of obsolescence and insufficient automation in this and other industries were fostered by shortages of capital. The latter was also leading to problems of energy supply because of the failure to build sufficient power station capacity. Industry was heavily concentrated into the largest cities, which had skewed industrial structures and which, because of the powerlessness of the local administrations relative to the big industrial ministries, suffered from inadequate investment in housing and social infrastructure. In the agricultural regions, the predominance of huge collective and state farms with relatively few settlements and roads led to difficulties in managing land utilization satisfactorily and increased the sense of rural isolation. This encouraged outmigration.

Soviet policies therefore left the region rather poorly prepared for the effects of economic transition. Problems of the oil and gas industries in the north were highlighted in chapter 5, but since 1991 they have been

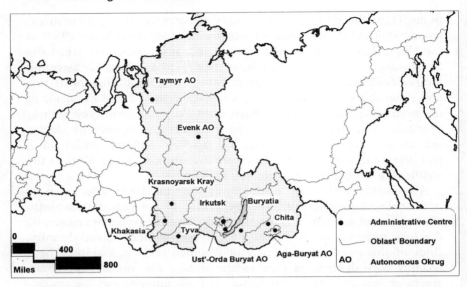

Figure 8.11   East Siberia economic region.

performing less badly than many other industries, and attracting some foreign involvement, at least at the margins. The significance of these industries to Russian exports is discussed elsewhere. By contrast, the coal industry of the Kuzbass is suffering cutbacks and is threatened with radical restructuring. Similarly, engineering has been contracting, to the detriment of cities like Novosibirsk, which has close ties to the defence industry. This city, however, has hopes of becoming Siberia's financial centre, even if earlier hopes of profiting from its record in science and technology (the city is home to the Siberian branch of the Russian Academy of Sciences) now appear hard to realize. While Siberians have long argued for a more balanced regional economy, which would tie the resource-producing north more closely to the manufacturing south, this seems unlikely in the foreseeable future. Market forces have if anything widened the gap between north and south, stimulated competition and divided Siberian regions into winners and losers.

The East Siberian Economic Region

East Siberia (figure 8.11) accounts for almost a quarter of Russian territory but only 6 per cent of its population (9.2 million people). The population is concentrated along the Trans-Siberian Railway to the south, where most of the cities are also to be found. A number of hydro-electric power sites further to the north are also locations for urban development. Apart from reindeer herding in the north, farming is mainly restricted to the south, with dairying around some of the big cities and arable and livestock farming in the steppe basins among the southern mountains. Much of East Siberia corresponds with the inhospitable Central Siberian Plateau, large parts of which are forested. To the north, however, the forests give way to tundra, as they do in areas of higher relief. Apart from the scattered settlements of the Siberian native peoples, the population of the north is limited to resource centres.

Russian penetration of East Siberia occurred in the seventeenth century in association with the fur trade. Thereafter, this often remote territory was valued above all for its natural resources. Russian settlers mixed with the native peoples, whose presence today is reflected in a network of republics and autonomous regions. Most of the north falls within Taymyr and Evenki autonomous okrugs (part of Krasnoyarsk Kray), while the three republics of Khakasia, Tyva and Buryatia are located in the south. In addition, Ust'-Orda Buryat and Aginsk-Buryat autonomous okrugs lie within Irkutsk and Chita oblasts respectively. The most heavily populated units, however, are the Russian-dominated ones: Krasnoyarsk Kray and Irkutsk and Chita oblasts.

Many of Leslie Dienes's comments regarding Soviet policy towards West Siberia apply with even more force to this region (Dienes, 1987). The development of a balanced and diversified regional economy was rendered impossible by the redirection of capital in the 1970s towards such prestige projects as the Baykal–Amur railway. Even under Brezhnev, Siberian manufacturing was accorded a reduced emphasis and the idea of locating energy-intensive industry to the east was only partially implemented. Under Gorbachev, who emphasized the revamping of industries in the European territories, even less cash was available for eastern development. Resource towns therefore suffer from narrowly based economies and from particular privations in social provision. Labour shortages have encouraged the use of temporary workers, especially in remoter regions.

Today, therefore, East Siberia's economy is focused on the processing of raw materials: chemicals and petrochemicals, non-ferrous metals and pulp and paper in particular. The availability of cheap electricity, the product of the Soviet emphasis on damming the great rivers of the region, is a distinct advantage. The region today has a number of major processing complexes which derive from the policies of the Soviet era: examples include the Noril'sk metallurgical combine in the far north, aluminium smelting in Bratsk, Krasnoyarsk and other centres and integrated pulp and paper complexes at Bratsk, Ust'-Il'imsk and elsewhere. Such resource-based activities have been performing relatively well in the new market conditions, seeking foreign partners and reorienting themselves towards export markets. As in other parts of Russia, however, official policy towards such strategic industries favours continued state involvement and privatization in favour of Russian investors. Other industries, like engineering with a military orientation, are not particularly focused on the regional economy and have not done well in the new circumstances.

The region's remoteness from both European Russia and the Pacific coast, and the rising costs of transportation, are obvious problems for the future. It may well be that the ambitious developments of the past may have to be reigned in, and those which do survive may be forced to become ever more dependent on outside interests. In a world of scarce resources, however, this region will surely continue to act as a supplier and processor of raw materials for Russia and an international market.

The Far East Economic Region

The enormous territory of the Russian Far East (figure 8.12) occupies over one-third of Russia (36.4 per cent), but contains only about 5 per cent of its population (7.8 million people). Most of the population is concentrated in the south, along the Trans-Siberian Railway, and in various coastal communities. The Far East is a land of great variation, including the great basin of the Lena and its tributaries, the numerous mountain chains to the south, east and northeast, fertile plains in the southeast and the peninsulas and islands of the Pacific littoral. Climatically, the region suffers from great continentality, but some what milder conditions are experienced in the south and near the coast, with the Pacific reducing the effects of continentality towards the east. Coastal regions and islands thus experience a maritime climate. While much of the territory lies within the boreal forest zone, tundra occupies the northern regions and the higher slopes of many mountain ranges. Pacific influences in the southeast give rise to a special form of mixed forest in some of the lowland regions. Apart from reindeer herding, agriculture is generally restricted to the southern basins with black earth soils, but even here it is hindered by drought, poor drainage, unseasonal frosts and other factors. Cultivation is also found far to the north around Yakutsk, and in sheltered parts of some of the islands and peninsulas.

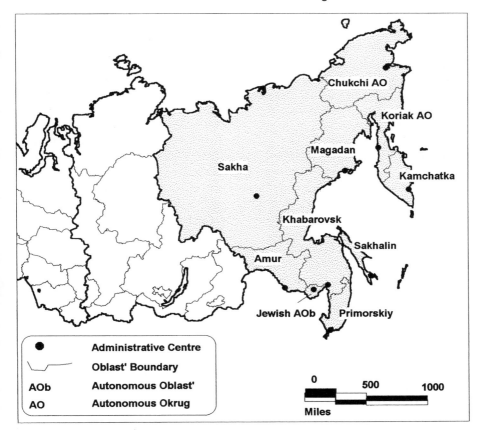

Figure 8.12   The Far East economic region.

Russian occupation of the Far East occurred in a number of phases: in the seventeenth century in association with the fur trade; in the following century when some of the outlying territories in the far northeast and the islands became sites for a maritime extension of the fur trade; and in the middle of the nineteenth century when territory along the Amur and in Primorskiy Kray was seized from China. The non-Russian peoples are today represented by a number of administrative units: the huge republic of Sakha (Yakutia) in the northwest, the Jewish Autonomous Okrug which is part of Khabarovsk Kray, the Koryak Autonomous Okrug, which is trying to gain independence from Kamchatka Oblast, and Chukchi Autonomous Okrug, which has successfully separated from Magadan Oblast.

Soviet priorities in the Far East were twofold: first, to exploit the region's rich endowment of natural resources; second, to take advantage of the territory's geographical location by developing its defensive and military capabilities and trying to seize export opportunities. The result was a very narrowly based economy which was heavily subsidized by Moscow and very much dependent on the centre for supplies and as the ultimate destination for most of its products. As with East and West Siberia, therefore, the economy was unbalanced and poorly coordinated regionally. As two writers suggest of Primorskiy Kray, and by implication most of the rest of this region, it was in a state of 'semi-colonial dependence' on Moscow and unable to take advantage of most foreign trade opportunities which might happen to exist (Kirkow and Hanson, 1994).

In the new situation of transition to the market, the region is suffering from a reduction in Moscow-based subsidies, rising costs of transportation for its products, a collapse of demand and the scaling down of the very considerable defence sector in this region. The scale of outmigration from northern regions was commented on in chapter 7. Notable activities in the Far East include the mining and processing of non-ferrous metals and minerals, forestry and food processing (notably fish). Unfortunately, quite apart from supply problems associated with high transportation costs (including problems of energy supply), the region's industries are in sore need of capital for re-equipping and restructuring. Military conversion is a particularly pressing issue. On the plus side, however, the Far East is relatively close to Japan, South Korea and China, some of the world's most dynamic economies in the recent past. It also has access to the Pacific basin as a whole, though its ports are in dire need of investment. However, the Far East economic region is so vast and varied that its potentials and problems differ greatly from one part to the next, and it is difficult to generalize.

Collapse of support from the centre has spurred regional leaders and entrepreneurs towards the taking of local initiatives, such as the setting up of special economic zones, encouraging partnerships and inward investment, cross-boundary commercial activities and, unfortunately, a good deal of criminal activity. Projects like the offshore oil and gas developments in Sakhalin, or gold mining in Magadan, clearly attract foreign interest, but where the risks are too high or the payback period is too long, foreign investors are more cautious. While the region's resources and geographical location seem to promise considerable potential for the future, it will obviously take many years before that potential can be fully realized, and even then many areas may remain too remote to benefit.

Nine

Russia's Autonomous Territories

Chapter 3 considered the establishment of the so-called 'autonomous territories' within the Russian Federation after 1917, and their developing relations with Moscow, especially after the tumultuous events of 1991. Here we will examine the autonomous territories both as a group and individually, in order to highlight their specific problems and aspirations. Chapter 8 claimed that it has never been more important to know about Russia's regions than now, in view of their varying characters, difficulties and new-found powers to determine their own futures. This is even more true of the autonomous territories.

Geographical Characteristics and Population of the Autonomous Territories

Russia's autonomous territories are for the most part situated on the periphery of the Federation (see figure 3.2). Collectively, they occupy an enormous area: no less than 53 per cent of the country's territory. In 1989, they accounted for about 17.6 per cent of Russia's total population. Population density varies enormously across the autonomous territories, from 0.4 persons per square kilometre in the Sakha Republic of the Far East (and even lower densities in some of the northern autonomous okrugs) to over 80 in part of the North Caucasus (table 9.1) (Harris, 1993b). The average for the Russian Federation as a whole in 1992 was 8.7. The northern autonomous territories are especially large and thinly peopled. Thus the two republics of Komi and Sakha, together with the seven autonomous okrugs belonging to the Peoples of the North, occupy 46 per cent of Russia's territory but contain only 3 per cent of its population.

As noted in chapter 3, the non-Russians constitute more than 18 per cent of the population of the Russian Federation, in excess of 27 million

Table 9.1 Autonomous territories: basic geographical characteristics

	Area (thousand sq. km)	Population, 1996 (thousand)	Population density (persons per sq. km)	Income per capita, 1993 (thousand rubles)	% Urban, 1996
Russia	17,075.4	147,976	8.7	884.7	73.0
North European/Middle Volga–Urals					
Karelia	172.4	785	4.6	937.8	74.0
Komi	415.9	1,185	2.8	1,337.2	74.4
Komi-Permyak AO	32.9	157	4.8	–	30.6
Mari-El	23.2	766	33.0	506.2	62.3
Mordovia	26.2	956	36.5	483.2	58.7
Chuvash Rep.	18.3	1,361	74.4	545.2	60.6
Tatarstan	68.0	3,760	55.3	807.3	73.5
Bashkortostan	143.6	4,097	28.5	968.3	64.7
Udmurtia	42.1	1,639	38.9	672.5	69.7
North Caucasus					
Kalmykia	76.1	319	4.2	377.8	38.6
Adygeya	7.6	450	59.2	294.2	53.8
Dagestan	50.3	2,098	41.7	206.4	41.8
Kabardino-Balkaria	12.5	790	63.2	306.1	57.7
Karachayevo-Cherkessia	14.1	436	30.9	358.7	46.1
N. Osetia	8.0	663	82.8	350.4	69.2
Chechnya		865			36.8[a]
Ingushetia		300		49.5	41.3
Siberia and the Far East					
Altay Rep	92.6	202	2.2	359.2	24.3
Buryatia	351.3	1,053	3.0	629.4	59.5
Tyva	170.5	309	1.8	265.4	48.5
Khakasia	61.9	586	9.5	908.9	71.7
Ust'-Orda Buryat AO	22.4	143	6.9	–	0.0
Aga-Buryat AO	19.0	79	4.2	–	32.9
Sakha Rep	3,103.4	1,023	0.33	2,130.0	64.3
Jewish AОb	36.0	210	5.8	570.5	67.1
Territories of the Far North					
Nenets AO	176.7	48	0.3	–	60.4
Khanty-Mansi AO	523.1	1,331	2.5	–	91.4
Yamalo-Nenets AO	750.3	488	0.7	–	83.0
Taymyr AO	862.1	47	0.06	–	66.0
Evenki AO	767.6	20	0.03	–	30.0
Chukchi AO	737.7	91	0.1	–	70.3
Koryak AO	301.5	33	0.1	–	24.2

Note: [a] Estimate.
Sources: *Rossiyskiy statisticheskiy yezhegodnik* (1994, pp. 6–8, 441–6), ibid. (1996, pp. 711–13), Bradshaw and Palacin (1996, pp. 43–60).

people. However, only about 52 per cent of the non-Russians live in the autonomous territories. In fact, Soviet policy towards the construction of ethnic homelands contained quite a number of contradictions (Dmitrieva, 1996, pp. 3–10). In the first place, the boundaries of the autonomous regions often failed to conform to the ethnic geography, leaving many of the titular nationality outside the homeland, but including numerous Russians and other nationality groups within it. Second, many groups were left without official homelands altogether. Thus in 1989, only 35 nationalities had official homelands, compared with over 90 numerically significant nationalities in Russia according to the census.

Chapter 3 noted that towards the end of the Soviet period a number of changes were made to the structure of autonomous units, and these have continued into the post-Soviet era. At the present time in Russia, there are 21 republics, one autonomous oblast (the Jewish) and ten autonomous okrugs. In the Soviet system, autonomous republics (ASSRs) had more rights than autonomous oblasts, and the latter more than the okrugs. In post-Soviet Russia all units, including the non-ethnic krays and oblasts, have similar rights according to the 1993 constitution, apart from a few symbolic ones. In practice, however, some units appear to be in a more privileged position than others. Some of the republics seem to be particularly favoured, especially the resource-rich ones. This issue was considered in chapter 3.

History of Settlement of the Autonomous Territories

From the geographical viewpoint (if not from the ethnic and the cultural), it makes sense to divide the autonomous territories into four groups (see figure 3.2). The first group extends right across the northern part of the Russian Federation, from the Finnish frontier in the west and then, with a break around the White Sea, eastwards to the Urals and on across northern Siberia and the Far East to the Pacific. A second group lies in a strategically important location around the middle part of the Volga River, in Volga–Vyatka and Volga regions, and stretching towards the Urals. The third group lies on the Russian side of the Caucasus mountains, bordering on Georgia and Azerbaijan, and stretching round the northwestern shore of the Caspian Sea almost to the delta of the Volga. The fourth group is in southern Siberia and the Far East, bordering on Mongolia and China.

These four groups can be related to Meinig's 'macrogeography of Western imperialism', discussed in chapter 1, since they form part of the story of Russian imperial expansion and the attempts to colonize those parts of the Russian state which lay beyond the original homeland. Thus the first group falls very much within what Meinig called the 'boreal riverine empire'. Here, we recall, Russian colonization took place in association with the quest for furs and the other natural resources of the boreal forest and tundra belts. Such settlement occurred from medieval times in north European Russia and essentially from the middle of the sixteenth century in the case of the regions lying east of the Urals. The native peoples were frequently obliged to pay tribute to the Russians, but often left to their traditional ways. West of the Urals, however, the degree of assimilation by the Russians was greater. Later in the nineteenth and twentieth centuries, Russian settlement intensified in association with resource-related developments.

The conquest of the second group of peoples of the middle Volga region and towards the Urals really began with Ivan the Terrible's seizure of the Tatar city of Kazan' in 1552. From here began the process of Russian settlement of the surrounding forest and forest-steppe lands. However, the native peoples, who had previously been subservient to the Tatars, continued to practise their agriculture, livestock farming, and hunting and gathering activities in those areas not settled by the Russians. The native peoples thus maintained their existence in this region, unlike many of the steppe regions, where pre-existing nomads were ultimately assimilated or displaced by Russian cossacks and agriculturalists. In Meinig's terminology, the whole area became part of the Russian 'settler empire'.

Russian contacts with the third group of peoples in the North Caucasus began in the sixteenth century but intensified in the late eighteenth, as the tide of Russian settlement began to approach the Caucasus Mountains and Russia began to extend its tentacles into the Trans-caucasian regions beyond. The mountain peoples of the Caucasus, however, put up a fierce resistance to the Russians, and it was not until the middle of the nineteenth century that they were pacified. Even after that, Russian influence was frequently minimal at the local level and sometimes challenged openly, most notably after 1917. This region could be said to form part of the 'nationalistic empire'.

Russian penetration of Siberia began in the sixteenth century, but the mountains and basins of its southern part and the adjacent Far East long remained an indeterminate area of contact between the Russians, the Mongols and the Chinese. Russian settlement of some parts intensified in the nineteenth century. The fourth group of autonomous territories therefore represents a meeting point between the 'settler' and 'nationalistic' empires.

The Russian presence in many of the autonomous territories greatly increased after 1917 as a result of such developments as the opening up of new resources, industrialization, collectivization, denomadization and the immigration of many Russian officials and administrators. The growth of the Russian population was particularly marked down to 1959 in most of the regions concerned. The main exception to this pattern was in the European territories outside the North Caucasus – Russians were already well represented in most of these by the time of the first Soviet census in 1926. After 1959, the increase in the proportion of Russians was less marked and even went into reverse in numerous cases. Lower rates of natural increase among Russian populations, especially compared with some of the indigenous groups in the North Caucasus and southern Siberia, was one reason. The return of previously exiled indigenous peoples to the North Caucasus, and the outmigration of Russians from there, seem to have been additional factors in that area. By contrast, across the north the proportion of Russians continued to grow even after 1959 in most cases, in association with resource developments. Altogether in 1989, out of 31 autonomous territories in Russia, the titular population had an absolute majority in only eight (table 9.2). Of course, since that date the proportion of Russians living in the autonomous territories will have fallen once again as a result of the mass outmigrations from the north and from such ethnic troublespots as Chechnya.

Economic Development of the Autonomous Territories

While the four groups of autonomous territories discussed above make sense from a geographical and also an historical perspective, they are far less satisfactory as a framework for considering cultural, political and current economic developments. The rest of this chapter will therefore employ groupings suggested by Chauncy Harris and others (Harris, 1993b), as follows (see tables 9.1 and 9.2):

1 North European and Middle Volga–Urals.
2 North Caucasus.
3 Siberia and the Far East.
4 Territories of the Far North.

Table 9.3 shows the basic employment structure of the 21 republics and ethnic regions for which data were available for 1990 and 1994.

Table 9.2 Autonomous territories: changing ethnic structures, 1926–89

	% total population titular, 1926	% Russian, 1926	% total population titular, 1989	% Russian, 1989
Russia	78	–	82	–
North European, Middle Volga–Urals				
Karelia	37	57	10	74
Komi	92	7	23	58
Komi-Permyak AO	77	23	60	36
Mari-El	51	44	43	47
Mordovia	36[a]	59[a]	33	61
Chuvash Rep	75	20	68	27
Tatarstan	45	43	48	43
Bashkortostan	23	40	22	39
Udmurtia	52	43	31	59
North Caucasus				
Kalmykia	76	11	45	38
Adygeya	45	26	22	68
Dagestan	59[b]	12	73[b]	9
Kabardino-Balkaria	76[c]	8	57[c]	32
Karachayevo-Cherkessia	56	3	41	42
N. Osetia	84	7	53	30
Chechen-Ingush	94[d]	3	71[d]	23
Siberia and the Far East				
Altay Rep	36	51	31	60
Buryatia	44	53	24	70
Tyva	57[a]	40[a]	64	32
Khakasia	50	47	11	79
Ust'-Orda Buryat AO	34[a]	56[a]	36	57
Aga-Buryat AO	48[a]	49[a]	54	43
Sakha Rep	82	10	33	50
Jewish AOb	9[a]	78[a]	4	83
Territories of the Far North				
Nenets AO	11[a]	69[a]	12	66
Khanty-Mansi AO	14[a,e]	72[a]	2[e]	66
Yamalo-Nenets AO	22[a]	45[a]	4	59
Taymyr AO	6[a,f]	65[a]	13[f]	67
Evenki AO	34[a]	58[a]	14	67
Chukchi AO	21[a]	61[a]	7	66
Koryak AO	19[a]	61[a]	16	62

Notes: [a] Data for 1959; [b] Avars, Dargins, Kumyks, Lezghins and Laks;
[c] Kabardinians and Balkars; [d] Chechens and Ingush; [e] Khanty and Mansi;
[f] Nentsy and Dolgany.
Source: Harris (1993b, pp. 554–6).

Table 9.3 Russia's autonomous territories: employment in major economic sectors, 1990–4

	Industry		Agriculture		Transport and communications		Construction		Trade		Services and otheres	
	1990	1994	1990	1994	1990	1994	1990	1994	1990	1994	1990	1994
Russia	30.3	27.1	12.9	15.0	7.7	7.8	12	9.9	7.8	9.5	29.3	30.7
North European, Middle Volga–Urals												
Karelia	33.4	29.4	5.6	5.5	10.2	12.0	11.4	9.6	8.9	10.3	30.5	33.2
Komi	28.5	28.3	5.5	5.9	10.4	11.2	17.0	12.9	8.9	9.7	29.7	32.0
Mari-El	31.9	28.1	17.4	20.9	5.2	4.9	11.8	8.0	6.8	8.3	26.9	29.8
Mordovia	31.2	28.8	19.2	19.6	6.5	6.7	11.6	8.1	6.7	7.6	24.8	29.0
Chuvash Rep	33.1	31.0	19.8	19.8	4.5	5.3	10.3	8.4	7.0	7.6	25.3	27.9
Tatarstan	31.9	29.3	13.6	14.7	7.2	7.1	13.4	11.6	7.0	8.4	26.9	28.9
Bashkortostan	29.9	27.7	16.3	20.0	6.3	6.3	12.9	11.0	7.2	7.3	27.4	27.7
Udmurtia	39.2	36.0	12.7	14.1	5.5	5.9	10.6	7.3	7.0	9.0	25.0	27.7
North Caucasus												
Kalmykia	11.4	10.1	30.0	33.8	6.7	6.2	14.2	7.7	6.9	7.4	30.4	34.8
Adygeya	–	28.0	–	15.7	–	7.6	–	6	–	7.8	–	34.9

Table 9.3 *Continued*

	Industry		Agriculture		Transport and communi-cations		Construction		Trade		Services and otheres	
	1990	1994	1990	1994	1990	1994	1990	1994	1990	1994	1990	1994
Dagestan	20.3	16.1	26.7	32.2	5.7	4.8	9.2	8.6	6.7	6.2	31.4	32.1
Kabardino-Balkaria	29.6	21.7	16.0	17.1	5.6	6.4	10.0	8.7	8.0	11.3	30.8	34.8
Karachayevo-Cherkessia	–	27.8	–	18.6	–	5.0	–	10.5	–	9.6	–	28.5
N. Osetia	32.3	27.0	11.0	12.0	5.8	5.7	9.6	10.5	8.4	8.6	32.9	36.2
Siberia and the Far East												
Altay Rep	–	9.2	–	32.6	–	4.7	–	7.1	–	7.4	–	39.0
Buryatia	23.9	21.8	14.4	16.7	9.1	9.0	13.2	9.1	8.3	9.4	31.1	34.0
Tyva	12.1	8.9	20.5	26.9	6.5	5.0	13.7	6.2	8.2	9.4	39.0	43.6
Khakasia	–	29.7	–	13.4	–	8.2	–	10.9	–	8.6	–	29.2
Sakha Rep	16.4	17.5	10.7	12.8	11.4	11.1	15.6	8.9	9.7	9.3	36.2	40.4
Jewish AOb	–	22.7	–	16.5	–	10.2	–	6.5	–	11.1	–	33.0
Territories of the Far North												
Chukchi AO	–	20.5	–	7.0	–	15.7	–	5.7	–	12.0	–	39.1

Source: Bradshaw and Palacin (1996, pp. 67–70).

Comparing the data for industrial employment in each region with the average for the Russian Federation as a whole, the North European (Karelia, Komi) and the Middle Volga–Urals regions (Mari-El, Mordovia, Chuvash Republic, Tatarstan, Bashkortostan, Udmurtia) all emerge as having above-average industrial employment in 1994. In the North Caucasus, however, this is true of only Adygeya and Karachayevo-Cherkessia, and of only Khakasia in Siberia. With respect to agricultural employment, a number of the Middle Volga–Urals regions (Mari-El, Mordovia, Chuvash Republic, Bashkortostan) and all of the North Caucasus ones apart from North Osetia have above-average agricultural employment. In Siberia the picture is rather more mixed, with above-average agricultural employment in four regions but not in Khakasia, while environmental conditions militate against agriculture in both Sakha and Chukotka.

As a generalization, it would be true to say that economic development in the autonomous territories has lagged behind that in Russia as a whole. Looking at the data on national income produced per capita in 1993 in table 9.1, it will be seen that this indicator exceeds the national average in only six regions (Karelia and Komi in North European Russia, Bashkortostan in the Urals, and Khakasia, Sakha and Chukotka in Siberia and the Far East). With the exception of Bashkortostan, all are resource-producing regions. Comparatively low levels of economic development characterize all the North Caucasian republics, as well as Altay and Tyva in Siberia. In most of these regions, the Russians are in a minority. The North Caucasian republics have been among the worst hit by the economic changes implemented since 1991.

Patterns of economic development and living standards across the USSR as they existed towards the end of the Soviet era have been analysed by Dmitrieva (Dmitrieva, 1996, pp. 65–97). According to her analysis, living standards were below the average for the USSR in every one of the 16 autonomous republics then existing in Russia. The group with the lowest living standards of all included four out of five autonomous republics in the North Caucasus and all three then existing east of the Urals (Buryatia, Tyva and Sakha). Patterns of economic development were more complex, with several of the North European autonomous republics having high levels despite their relatively low living standards. Once again, the North Caucasian and Siberian republics lagged behind, with the exception of Sakha, which had a highly subsidized form of development.

The economic development of the various republics and ethnic regions during the Soviet period has a bearing on the extent to which they have been able to retain their ethnic distinctiveness (table 9.2). Resource-based developments in Karelia and Komi, particularly during and after the Second World War, led to an influx of Russians. This was especially the case in Komi, which had an overwhelming majority Komi population in 1926 reduced to less than a quarter by 1989. Resource-based developments also affected republics and regions in the Middle Volga–Urals area, especially after the Second World War. The development of manufacturing industry here tended to lag behind that of the neighbouring non-ethnic regions: for example, in Volga–Vyatka. There was a deliberate policy to even out development from the 1960s, when the authorities began to worry about labour shortages, but only a moderate influx of Russians occurred in most of the autonomous territories of this area. The North Caucasus experienced considerable influxes of Russians during the Soviet period in association with agricultural, energy and mineral-based developments as well as with the temporary exile of some of the native peoples. More recently, as noted above, reproduction rates among the indigenes have often outpaced those of the Russians, and there has been outmigration among the latter. Resource-oriented development attracted Russian migration into some of the Siberian regions (notably the hydro-electric power oriented industrial complex in Khakasia and various developments in Buryatia) and across the north, including Sakha. Fewer Russians, however, were attracted to the less developed regions like Tyva.

With the exception of the Middle Volga–Urals area, therefore, economic development has tended to be resource-oriented, though agriculture remains important in many cases. Soviet economic development policies had the effect of reducing the ethnic distinctiveness of many autonomous regions. Since 1991, the autonomous regions of Russia, like other Russian regions, have had to come to terms with their new post-Soviet situation and to face up to the consequences of their Soviet heritages. They have also had to take more responsibility for their own futures. Different republics and regions have been seeking varied answers to their problems, depending upon their individual circumstances. Such circumstances will now be considered as we turn to examine the republics and regions according to their designated groups.

North European and Middle Volga–Urals Group

This is arguably the most geographically diverse group, consisting of territories which have been associated with the Russian state from at least the sixteenth century (table 9.4). Harris has subdivided the group

Table 9.4 Ethnic characteristics of the North European and Middle Volga–Urals group

	1	2	3	4	5
Karelia	785	10	74	51.5	63
Komi	1,185	23	58	74.3	87
Komi-Permyak AO	157	60	36	82.9	65
Mari-El	766	43	47	88.4	50
Mordovia	956	33	61	88.4	29
Chuvash Rep.	1,361	68	27	85.0	51
Tatarstan	3,760	48	43	96.6	32
Bashkortostan	4,097	22	39	74.7	64
Udmurtia	1,639	31	59	75.7	69

Key: 1, total population, 1996; 2, % titular population, 1989; 3, % population Russian, 1989; 4, % titular population in homeland speaking native tongue as first language, 1989; 5, % titular population residing in homeland, 1989.
Source: 1989 Census of Nationality, Harris (1993b), *Rossiyskiy statisticheskiy yezhegodnik* (1996, pp. 711–16).

into a North European part (republics of Karelia and Komi, together with the Komi-Permyak Autonomous Okrug) and a Middle Volga–Urals part (republics of Mordovia, Chuvash, Mari-El, Tatarstan, Udmurtia and Bashkortostan). The former lie in the Northern Economic Region, largely in the boreal forest, with poor agricultural potential and low population densities (figures 9.1 and 9.2). Their development has been mainly resource-oriented. The latter are in the well populated mixed-forest and forest-steppe regions of Russia, with a more variegated development based upon extractive industries, manufacturing and agriculture. Ethnically and linguistically, the Karelians, Komi, Komi-Permyaks, Mordvinians, Mari and Udmurts are Finno-Ugrians. The Tatars, Bashkirs and Chuvash are generally regarded as Turkic peoples, although they have had diverse ethnic histories, especially the Chuvash.

The resource-oriented economies of both *Karelia* and *Komi* are well illustrated by table 9.5, with Karelia specializing in electro-energy, ferrous metals and forestry industries and Komi in fuels and forestry. The republics' long association with Russia, together with their development patterns in the Soviet era, have meant an influx of Russian settlers, and both indigenous peoples are easily outnumbered in their own republics. The Karelians especially seem to have travelled a long way down the

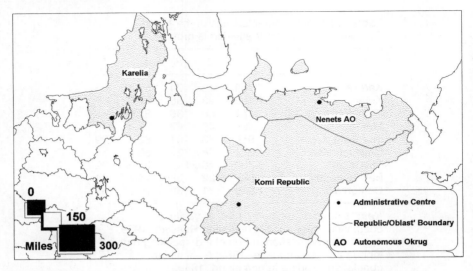

Figure 9.1   Autonomous territories of north European Russia (excluding Komi-Permyak AO).

road of integration with the Russians, since their absolute numbers have been falling (through assimilation and to some degree low reproduction rates) and only 63 per cent of Karelians lived in Karelia in 1989. Moreover, even in the republic only 52 per cent of Karelians spoke Karelian as their first language. More of the Komi reside in their own republic and they are more faithful to their mother tongue, but they too have been swamped by Russians, especially in the Soviet period. The fact that both the Karelians and the Komi were converted to Russian Orthodoxy at an early period may have reduced their propensity for asserting their cultural distinctiveness and differences from the Russians. Policy towards Moscow has tended to be non-confrontational, but there is official support for cultural development: for example, of the Komi language.

The *Komi-Permyak Autonomous Okrug* (figure 9.2) is relatively underdeveloped and predominantly rural. The region lacks resources other than timber and has no rail link to the outside world. With a total population of only 160,000, it is obviously in no position to assert its independence. Unsurprisingly, the indigenous population easily outnumbers the Russians, but there seems to have been considerable assimilation into the Russian population over the years.

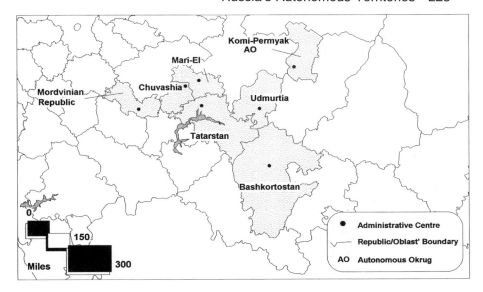

Figure 9.2   Autonomous territories in the Middle Volga–Urals region (including Komi-Permyak AO).

The remaining six republics in this group lie in a strategic location across the major routes between central Russia and the east. They form a solid block of autonomous territory overlapping the Volga–Vyatka, Volga and Urals economic regions (see figure 9.2 and chapter 8). Historically, the entire region was a meeting point between Finno-Ugrian, Turkic and Slav peoples, who approached it from different directions: hence the ethnic mixing which characterizes it today. Once it was absorbed by the Russian state, however, the region was subjected to considerable Russian immigration, and today it is surrounded by a sea of Russian settlement. This means that it would be difficult if not impossible for these republics to secede from the Russian Federation. Equally, any attempted secession would be extremely worrying to Moscow because of the republics' strategic location.

Before the Middle Volga was annexed by Russia, the region was dominated by the *Tatars*, whose power was focused around the formidable Khanate of Kazan'. This state of affairs was brought to an end by Ivan the Terrible in 1552, when he captured Kazan' and overthrew the khanate. However, the predominantly Islamic Tatars remained a culturally significant force and in the nineteenth century various nationalistic

Table 9.5 North European and Middle Volga-Urals group: sectoral structure of industrial production, 1994 (% of total regional production)

	1	2	3	4	5	6	7	8	9	10	11	12
Russia	13.4	15.1	8.3	6.6	7.4	19.6	4.6	4.8	0.4	3.1	12.5	2.2
Karelia	13.5	–	14.9	4.3	0.1	7.0	40.8	4.3	–	0.8	11.3	2.5
Komi	16.1	50.6	0.1	0.1	–	2.9	16.6	4.3	–	1.6	5.8	0.7
Komi-Permyak AO	6.7	–	–	–	–	12.0	39.1	6.3	–	1.8	26.4	3.8
Mari-El	20.4	0.2	0.1	–	5.9	25.5	10.2	7.4	–	4.4	17.3	4.0
Mordovia	12.4	–	0.2	–	10.8	45.1	1.0	7.5	–	3.0	16.3	3.1
Chuvash Rep.	13.4	0.1	0.2	–	12.2	34.4	2.7	6.8	–	9.7	15.5	3.6
Tatarstan	12.3	19.4	0.5	–	22.3	22.5	1.7	3.7	0.1	3.5	10.6	2.1
Bashkortostan	11.1	30.2	1.7	1.8	20.5	13.0	1.7	3.7	0.8	2.4	9.4	2.3
Udmurtia	12.8	11.1	6.0	–	0.8	44.3	4.0	3.0	1.0	1.4	10.9	3.8

Key: 1, electro-energy; 2, fuels; 3, ferrous metals; 4, non-ferrous metals; 5, chemicals and petrochemicals; 6, metalworking and machine building; 7, forestry, woodworking, pulp and paper; 8, building materials; 9, glass and china; 10, light industries; 11, food industries; 12, milling and feed.
Source: Sravnitel'nyye pokazateli ekonomicheskogo polozheniya regionov Rossiyskoy Federatsii (1995).

and religious movements sprang up among them. Pan-Islamic and Pan-Turkic sentiments were among those which took root among Tatar intellectuals and others. At the time of the October Revolution, some Tatar leaders and intellectuals hoped that the new Bolshevik government would be sympathetic to the establishment of an Idel Ural state to incorporate all the indigenous peoples of the Middle Volga–Ural region under Tatar suzerainty. These hopes were dashed by the setting up of a separate Bashkir ASSR in March 1919, followed by the Tatar ASSR just over a year later. The Bolsheviks were, of course, suspicious of both religion and nationalism, and were particularly unhappy about any national movement which might jeopardize Soviet unity. The fact that the new Tatar republic contained only about a quarter of all Tatars living in the region, and that the Tatars were the dominant ethnic group in Bashkiria, speaks volumes about their motivation for refusing to countenance Tatar ambitions. Later manifestations of Tatar nationalism were ruthlessly suppressed by Stalin, who refused to consider the upgrading of the Tatar ASSR to Union republic status (despite its significant population) on the grounds that it was completely surrounded by Soviet territory. This was justified by the fiction that it could not exercise its constitutional right to secede from the USSR should it wish to do so (however unlikely this was under Soviet conditions).

The Tatar Republic already had a significant Russian population in 1926, and this has more or less maintained itself ever since. In 1989, the Tatars had 48 per cent of the total population and the Russians 43 per cent. The republic's economy grew rapidly after the outbreak of the Second World War, and especially with the oil- and gas-related developments and associated industrial growth of the Volga region in the postwar period (see chapter 8). As well as the major city of Kazan', with over a million people, the republic contains the important KamAZ truck plant at Naberezhnyye Chelny. In the mid-1990s, its oil production amounted to about 8 per cent of the Russian total. The republic's economic profile reflects the importance of the fuels, chemical, petrochemical and engineering sectors (table 9.4), and its oil resources have helped to encourage claims for more autonomy.

Given the national history of the Tatars, careful observers were not surprised when Tatar nationalism began to reassert itself in the wake of other national movements of the late 1980s. Longstanding grievances included the issue of the political status of *Tatarstan* (as the republic was soon to be renamed), its control over its own economy and resources, including sources of revenue, and the problem of the many Tatars who lived beyond the borders of the republic. In August 1990, the republic declared its sovereignty in a document which made no mention of its being part of the Russian Federation. Soon it established its own presidency and refused to sign the Federation Treaty which was negotiated between the federal government and 18 of its then 20 republics in March 1992. In the same month, a referendum in Tatarstan posed the question of whether the voters considered the republic to be a 'sovereign state' and a 'subject of international law' which was 'entitled to develop relations with the Russian Federation and other states on the basis of treaties between equal partners'. This virtually amounted to a bid for independence. The referendum vote proved a positive one, attracting the support of many Russians as well as Tatars. Similar sentiments were expressed in the new constitution, approved in November 1992, while only a small minority of Tatarstan's voters participated in the referendum on the Russian constitution in December 1993.

Economic pressure on Tatarstan was increased following Yeltsin's victory over the Russian parliament in October 1993. Finally, in February 1994, the republic was induced to agree to a bilateral treaty with the federal government which defined Tatarstan's status as a 'state united with Russia'. Many other aspects of the relationship between the two governments were more carefully defined at the time, but there are numerous loopholes, including the legal problem of reconciling the treaty with both the Tatarstan and the Russian constitutions (Magnusson, 1996). Whether Tatarstan will remain content with its present status within Russia, or yet push for further concessions, remains to be seen. Since Tatars live well beyond the boundaries of Tatarstan, and there is a long history of Tatar cultural influence over neighbouring peoples, any future resurgence of Tatar nationalism or Islamic militancy has the potential to spread over a significant area of central and southern Russia.

Bashkortostan, lying to the east of Tatarstan, is another republic which has taken a rather independent stance. The nominally Islamic Bashkirs are descended from nomadic peoples who were only slowly subjected to Russian control from the seventeenth century onwards. They have developed under the cultural influence of the Tatars – in 1989 the Bashkirs, who constituted only 22 per cent of their republic's population, were outnumbered by both the Russians (with 39 per cent) and the Tatars (with 28 per cent). In fact, out of 864,000 Bashkirs residing in their titular republic, only about 75 per cent spoke the mother tongue as their first language, while 179,000 spoke Tatar and another 40,000 Russian. Altogether, almost twice as many residents of Bashkortostan spoke Tatar as their first language as Bashkir. It may be the cultural influence of Tatars in Bashkortostan, and the example of Tatarstan, which helped to encourage Bashkir nationalism, even though the republic is swamped by non-Bashkirs and contains only 64 per cent of all Bashkirs in Russia. There has in fact been relatively little evidence of a history of ethnic pride among the Bashkirs (Todres, 1995). Issues behind the drive towards greater autonomy included the usual ones of control over resources and property, income from foreign trade (especially resource exports) and the distribution of taxation between the republic and Moscow. As in many other republics, the drive was sponsored by a local elite who were anxious to hold on to their power in the new circumstances and to increase their wealth if at all possible. Bashkortostan made a declaration of sovereignty in October 1990. It was induced to sign the Federation Treaty in March 1992 only after it had secured an additional protocol enhancing its control over resources and its taxation rights. The Bashkir constitution of December 1993 described Bashkortostan as a 'sovereign state' and declared that 'the relations between Bashkortostan and the Russian Federation are based on treaties'. Such claims are contrary to the Russian constitution. Finally, Bashkortostan signed a bilateral treaty with Russia in August 1994, which recognized its rights to natural resources and the oil refining and petrochemical plants, while Moscow retained control over the much less valuable coal industry and military complex.

Bashkortostan's industrial structure shows a bias towards the fuels, chemical and petrochemical sectors (table 9.4). The republic's development was very much linked to the rise of the Volga–Urals oil and gas field and to the building of oil and gas pipelines through the territory. Although oil production peaked many years ago, the republic is able to trade on its key position as a centre for refining and for petrochemical and chemical industries.

The remaining four republics of the Middle Volga–Urals region (*Mari-El, Mordovia, Chuvash Republic* and *Udmurtia*) have their own distinctive characters, but none has pursued the confrontational course of Tatarstan and Bashkortostan. As noted in chapter 8 (see the sections on Volga–Vyatka and the Urals), all are characterized by significant development of the machine building and engineering sectors, and have played an important role in the military–industrial complex (especially Udmurtia). All have suffered severely in the current transition. The Russians constitute a minority in the ethnically Hunnic-Turkic Chuvash Republic, but in 1989 they outnumbered the native populations in the other three (barely so in Mari-El). All four republics also have a minority Tatar population. The percentage of the titular population residing in the home republic varies from 69 per cent in the case of the Udmurts to only 29 per cent in that of the Mordvinians. All this is testimony to the ethnic and cultural mixture of the entire region, with ethnic groups being quite dispersed and intermixed. In fact, the situation is even more complicated than might be imagined at first, since both the Mari and the Mordvinians are divided into two groups speaking mutually incomprehensible dialects. Religiously, too, there is diversity, with Islam threatening to make inroads among the Orthodox Chuvash and Mari, and the latter together with other Finnic peoples retaining elements of pre-Christian animist and shamanist belief. It has been claimed that it is the refusal of the Mari to adopt either Orthodox or Islamic cultures which has helped them to preserve their ethnic identity. However, all four peoples have been experiencing a decline in adherence to their native languages even in their own republics. This was fomented during the Soviet period by policies which discriminated against their use in education and for official purposes. There has also been a reduction in rates of population growth over a lengthy period. All this suggests assimilation in favour of the majority Russian population, with the Mordvinians being especially affected.

The North Caucasus

The North Caucasus group consists of the eight republics of *Adygeya, Karachayevo-Cherkessia, Kabardino-Balkaria, North Osetia, Ingushetia, Chechnya, Dagestan* and *Kalmykia-Khal'mg-Tangch* (figure 9.3). The first seven fall in the North Caucasus Economic Region, while Kalmykia is part of the Volga Region. Both Adygeya and Karachayevo-Cherkessia were autonomous oblasts during the Soviet period but have since been upgraded to republics. The Chechen and Ingush republics were formerly a single republic but split up in 1992.

There is no doubt that the North Caucasus group of republics represents the most formidable challenge to date to the territorial unity of the Russian Federation. With the exception of Kalmykia, which is the historic home of the formerly nomadic Kalmyk peoples who once wandered the steppes in frequent competition with the Tatars, the republics lie in the northern foothills of the Caucasus mountains and extend outwards into the neighbouring parts of the steppe. Here the Russian, Ukrainian and cossack settlement of the 'settler empire' came into conflict with the stout resistance of the mountain peoples. Most of these peoples, apart from the majority of the Osetians, are Islamic in culture, and this reinforced their determination to fight the Russian intruder. As noted above, it was not until the middle of the nineteenth century that the mountain peoples were quelled, and even then Russian control was only partial. The region thus falls within what Meinig called the 'nationalistic empire' (see chapter 1).

Ethnically and linguistically, the peoples of this area fall into a bewil-

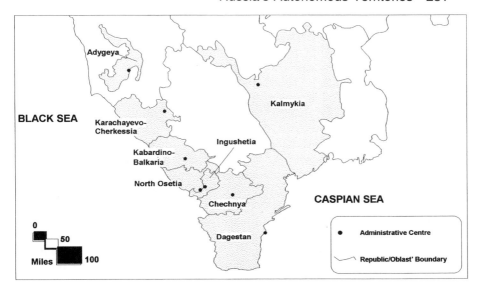

Figure 9.3   Autonomous territories of the North Caucasus.

dering variety of groups, including the Ibero-Caucasians (Adygeyans, Kabardians, Cherkessians, Chechens, Ingush, several of the peoples of Dagestan), the Indo-Europeans (notably the Osetians), the Altaic peoples (Karachay, Balkars) and the Mongols (Kalmyks). The fact that the region has long been a meeting point between peoples, coupled with the mountainous terrain which helped to preserve languages, dialects and ethnic sub-groups, seems to have been among the major causes of this cultural admixture. Down to the twentieth century, moreover, the peoples of the region felt at least as much loyalty towards their clan or tribe as they did towards their ethnic group. Equally, however, similarities across the region in numerous economic and cultural traits meant that there was a certain loyalty towards a 'mountaineer' identity which encouraged cooperation and mutual alliances on many occasions, notably in the nineteenth-century Caucasian wars against the Russians and in the establishment of a Mountain Republic in the years after the 1917 Revolution.

Perhaps because of their suspicions of the independent spirit of the North Caucasians, the Bolsheviks refused to countenance their incorporation into a single republic and soon began to split them up and then to join them together into units which did not conform to their linguistic and ethnic divisions. No doubt this was to make real unity more difficult, on the principle of 'divide and rule'. Thus the Kabardians were grouped with the Balkars, despite the fact that the former have more in common with the Cherkessians and the latter with the Karachay. Dagestan, by contrast, lacks a titular ethnic group, since it groups together at least ten different peoples. What made matters worse was the exile by Stalin in 1944 of five of the region's ethnic groups, namely the Karachay, Balkars, Chechens, Ingush and Kalmyks. The peoples were sent to Central Asia and other regions in punishment for their alleged collaboration with the Germans in the Second World War. Their ethnic units were abolished and their lands reallocated to others. The exiles were finally permitted to return after 1956, only to find that their former homes and lands were occupied by other people. This has caused many problems, especially among the Ingush, who have significant territorial claims against and grievances towards the Osetians.

In 1989, the Russians outnumbered the titular peoples in only Adygeya and narrowly in Karachayevo-Cherkessia (table 9.6). As noted above, Russians were attracted into the region by the economic developments of the Soviet period, which included oil production, refining and related activities (including petrochemicals) around Groznyy in the then Chechen-Ingush Republic, oil production around Maykop in Adygeya and also in Dagestan, hydro-electric power in various places in the Caucasus, the production and processing of metallic ores in North Osetia, Kabardino-Balkaria and Karachayevo-Cherkessia and building materials production (table 9.7). Agriculture and food processing have also been important activities. We have already commented on the relative underdevelopment of most of the North Caucasian republics and on the relatively low living standards. These have naturally been

Table 9.6 Ethnic characteristics of the North Caucasus group

	1	2	3	4	5
Kalmykia	319	45	38	96.1	88
Adygeya	450	22	68	98.4	78
Dagestan	2,098	73	9	84.4[a]	79–91[b]
Kabardino-Balkaria	790	57	32	98.8[c]	94/90[c]
Karachayevo-Cherkessia	436	41	42	98.9[c]	86/79[c]
N. Osetia	663	53	30	98.2	83
Chechen-Ingush	1,165	71	23	99.7[c]	82/76[c]

Key: see table 9.4.
Notes: [a] seven peoples; [b] range for seven peoples; [c] two peoples.
Source: As for table 9.4.

Table 9.7 North Caucasus group: sectoral structure of industrial production, 1994 (% of total regional production)

	1	2	3	4	5	6	7	8	9	10	11	12
Russia	13.4	15.1	8.3	6.6	7.4	19.6	4.6	4.8	0.4	3.1	12.5	2.2
Kalmykia	33.0	5.8	–	–	–	6.9	0.8	9.1	–	1.7	30.8	5.4
Adygeya	3.5	1.1	–	–	0.1	13.7	15.9	9.4	–	3.7	46.4	4.2
Dagestan	19.8	4.9	–	–	2.3	18.7	0.9	12.8	0.4	2.3	27.3	9.4
Kabardino-Balkaria	13.4	–	0.1	9.6	0.3	26.6	1.5	7.0	0.2	7.9	19.3	4.5
Karachayevo-Cherkessia	11.6	0.3	–	3.1	22.4	13.2	2.3	21.0	0.1	3.4	15.9	2.5
N. Osetia	10.1	–	–	19.2	2.0	15.2	2.8	7.7	2.2	6.1	22.1	11.1

Key: see table 9.5.
Source: as for table 9.5.

much affected by the ethnic strife and turbulence of the post-1991 period.

The period since 1991 has seen an outflow of Russians and other refugees from the North Caucasus republics in the wake of the political problems and violence which have afflicted many areas. Nowhere has the violence been worse than in *Chechnya*. Problems in the Chechen Republic date both from the circumstances surrounding the 1944 deportation and its aftermath, and from the republic's declining economy during the latter part of the Soviet period (Bond and Sagers, 1991). Historically, the Chechens were among the most warlike of the Caucasian peoples, with a strong commitment to Islam and particularly determined to resist Russian imperial expansion in the nineteenth century. In the later Soviet period, the Chechen-Ingush Republic was split ethnically (notably, between the Chechens, the Ingush and the Russians) and also politically among the Chechens themselves. The disputed election of Dzhakar Dudayev to the republic's presidency in October 1991 heralded a confrontational policy towards the Russian Federation. Within a month, Chechnya had declared itself independent and Yeltsin's attempt to take forceful action was frustrated by the Russian parliament. Ingushetia decided to separate from Chechnya in 1992. Subsequent attempts by the Yeltsin government to come to a compromise with Dudayev proved abortive – Chechnya, for example, refused to sign the Federation Treaty in 1992 or to take part in subsequent elections. Meanwhile, the Russians manoeuvred to set up a rival government. Finally, in December 1994, the decision was taken to invade the breakaway republic, possibly the result of pressure by the Russian army. Despite the terrible suffering of the Chechen population and mass migrations away from the scene of the fighting, Chechen resistance continued and included actions taken against Russian targets outside Chechnya itself. Dudayev himself was killed in the fighting. Protracted negotiations for a peace settlement continued to 1997.

Other tensions in the region have included the violence between North Osetians and Ingush over territorial claims, the struggle by the South Osetians of Georgia to unite with their northern brethren and attempts to establish separate ethnic republics in Karachayevo-Cherkessia, Kabardino-Balkaria and for some of the ethnic groups in Dagestan. It remains to be seen whether the political geography of the area will change further in the future. Clearly, unlike the case of republics in other parts of the Russian Federation, secession by one or more of those in the North Caucasus would hardly pose a mortal threat to the Federation itself. Public opinion surveys in Russia seem to indicate no strong support for forcibly keeping the Chechens within the

Russian state. The problem for Russian politicians to ponder is whether letting the Chechens go is setting too dangerous a precedent. The problem for the region's nationalities is to decide whether or not it is in their interests to remain with Russia, and how best to live with their neighbours.

Siberia and the Far East

Strung out along eastern Russia's southern border with Mongolia is a series of autonomous territories: the republics of Altay, Khakasia, Tyva and Buryatia, and the autonomous okrugs of Ust'-Orda-Buryatia and Aga-Buryatia. The Jewish (Yevreyskiy) Autonomous Oblast lies further to the east on the Amur River border with China (figure 9.4). Also coming within this group of autonomous territories is the Sakha (Yakut) Republic, which, although located in the far north, contains a numerous titular population whose aspirations and problems differ substantially from those of their neighbours, the smaller peoples of the north. The latter will be considered in the next section.

The titular nationalities of this group of autonomous territories belong either to the Turkic (Altay, Khakas, Tyvans, Sakha) or to the

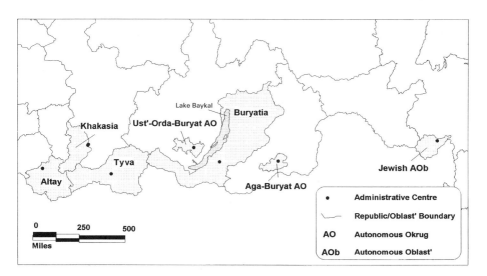

Figure 9.4   Autonomous territories of southern Siberia and the southern Far East.

Mongol (Buryats) peoples. The Jewish Autonomous Oblast is an exception, in the sense that it was an artificial creation, officially designated in 1934 as a homeland for the Soviet Jews who had no historic settlement status in the region. Most Jews live in the western part of Russia, especially in cities, and many have now emigrated to Israel, the United States and other lands. It was originally hoped that the region might provide a counterweight to the attractions of Palestine as a potential Jewish homeland. Despite Soviet pressure in the 1930s, few Jews moved there and any notion of developing a genuine Jewish culture in the territory was eventually undermined by Stalinist anti-Semitism. In 1989, only about 9,000 Jews lived in the area, out of a total population at that time of 214,000. More than 80 per cent of the population is Russian.

As noted above, the southern autonomous territories lie in a mountainous region which was an historic borderland between Russian, Chinese and Mongol influences. The native peoples were by tradition mainly nomadic herders, but their lands were gradually subjected to pressure from Russian settlers who moved in to take up farming, mining and other activities. In Altay, Khakasia and Buryatia the Russians already had a substantial presence by 1917. Subsequent immigration by Russians and others has been encouraged by the economic developments of the Soviet period. Today the Russians are outnumbered by the native populations only in Tyva and Aga-Buryatia (table 9.8). In remote Sakha, the Russians were outnumbered by the Sakha until the 1960s. Immigration by non-natives was stimulated through resource exploitation, and by 1989 the Sakha formed only about one-third of the population of the republic. The economic orientation of

Table 9.8   Ethnic characteristics of the Siberia and Far East group

	1	2	3	4	5
Altay Rep	202	31	60	89.6	85
Buryatia	1,053	24	70	89.4	82
Tyva	309	64	32	99.0	96
Khakasia	586	11	79	83.2	80
Ust'-Orda Buryat AO	143	36	57	90.0	—[a]
Aga-Buryat AO	79	55	41	98.4	—[a]
Sakha Rep	1,023	33	50	95.1	96
Jewish AOb	210	4	83	11.7	2

Key: see table 9.4.
Notes: [a] see entry for Buryatia.
Source: as for table 9.4.

the various territories is apparent from tables 9.3 and 9.9, with most of the southern ones having significant agricultural and food processing sectors. Extractive industries and raw materials processing are generally important. Only the Jewish Autonomous Oblast exceeds the Russian Federation average for engineering, but light industries are more widespread.

As in other parts of the USSR, Soviet policy worked against unity between what were essentially still tribal peoples in 1917, splitting them up between different territories. Thus the Altay, the Khakasy and their near neighbours the Shors (whose autonomous okrug was disbanded in 1939, with the area being added to Kemerovo Oblast) had relatively little national consciousness by 1917. Nevertheless, there was a move to form a pan-Turkic republic, but this was forestalled by the Soviets, who insisted on establishing separate national areas. Ironically, the development of a sense of national identity among these peoples can thus be credited to the Soviets. Buryat nationalists evinced a desire for unity with their Mongolian cousins south of the border, but once again this was frustrated by the Soviets. In the 1920s, the Mongol-Buryat ASSR was established to embrace all the Buryat lands in the USSR, but Soviet linguistic and cultural policy was devoted to emphasizing the differences between the Buryats and the Mongols. Finally, in 1937, after a purge of Buryat intellectuals, Buryat territory was divided into the three presently existing units and part of it was given to Irkutsk and Chita oblasts. The Buryat-Mongol ASSR was renamed the Buryat ASSR in 1958.

We shall now briefly consider each of the Siberian/Far Eastern republics in turn.

Altay Republic (formerly Gorno-Altay Autonomous Oblast) This republic is situated in the West Siberian economic region. In 1990, it assumed republican status and thus separated itself from Altay Kray, where the Russians are overwhelmingly predominant. The Altay people consist of two major groups with distinct languages and of several different sub-nationalities within each group. There has been some tension between these different peoples, and thus the process of nation building around a single Altay identity is incomplete. Policy has favoured the preservation of the Altay languages and culture and, since the republic is largely agricultural and rural and most Altay live in the countryside and speak their native languages, the preservation of Altay identity seems assured for the foreseeable future. The 1990 sovereignty declaration proclaimed republican ownership of its natural resources, but the most significant resources in this republic are agricultural ones.

Table 9.9 Siberia and Far East group: sectoral structure of industrial production, 1994 (% of total regional production)

	1	2	3	4	5	6	7	8	9	10	11	12
Russia	13.4	15.1	8.3	6.6	7.4	19.6	4.6	4.8	0.4	3.1	12.5	2.2
Altay Rep	0.8	–	–	9.0	0.2	7.8	10.9	12.4	–	15.0	40.8	0.4
Buryatia	32.1	6.4	0.7	8.6	–	17.3	7.5	6.1	0.3	3.0	13.8	3.0
Tyva	12.6	15.0	–	17.7	0.1	5.0	3.7	6.7	–	3.5	25.8	7.1
Khakasia	17.2	4.7	–	45.3	0.1	4.7	1.3	5.7	–	5.7	7.1	1.6
Ust'-Orda Buryat AO	–	5.6	–	–	–	11.8	19.8	6.2	–	4.1	28.8	23.2
Aga-Buryat AO	65.8	0.5	–	6.7	–	2.3	3.6	2.5	–	0.2	14.2	0.1
Sakha Rep	12.0	15.6	–	58.4	–	1.4	1.4	5.6	–	0.3	4.5	–
Jewish AOb	20.4	–	–	47.6	0.1	0.4	0.6	1.3	–	0.2	4.2	–

Key: see table 9.5.
Source: as for table 9.5.

Khakas Republic (formerly Khakas Autonomous Oblast)   Like the Altay, the Khakasy were originally subdivided into a series of tribes and regional groups, which were united by the Soviets into a single homeland. Unlike the Altay Republic, however, Khakasia, which is in East Siberia, has been subjected to intensive industrialization in association with the development of the Sayan Territorial Production Complex along the upper Yenisey River. This involved the damming of the river and its exploitation for hydro-electric power, together with coal mining, aluminium smelting and heavy engineering. Ethnically this resulted in a dramatic increase in the Russian population, and only 11 per cent of the population were ethnic Khakasy in 1989. However, as in many other autonomous regions, the two groups are spatially separate. Seventy-eight per cent of the Russians in 1989 lived in urban areas, but only 36 per cent of the Khakasy. There is a relatively high retention of the native language among the Khakasy (77 per cent on average but 84 per cent for those living in Khakasia). Even so, this level is the lowest for the Siberian/Far Eastern group of autonomous territories (the exception, of course, is the Jewish Autonomous Oblast, where Russian is the main language even among the Jews). One of the main expressions of Khakasian national feeling in recent years has been the Siberian Cultural Centre, founded in 1989 in association with the Altay and the Shors. This has been used to promote regional and ethnic causes, such as greater autonomy, native language retention, environmental protection and inter-ethnic cooperation.

Tyva Republic (formerly the Tuva ASSR)   Tyva is one of the least developed of Russia's republics, with a predominantly rural population living a traditional, largely pastoral lifestyle. The Russians are a minority, though many were attracted into Tyva during the Soviet period because of mining developments, including non-ferrous metals and minerals. However, more recently Russians have been leaving the republic in considerable numbers in the wake of ethnic violence directed against them. One factor behind this seems to be the relatively recent incorporation of Tyva into the USSR, in 1944, before which the republic was nominally independent. This means that Tyvans have a recent memory of existence as an independent state. There has been considerable resentment at the effects of Soviet policy, which included attempted suppression of the predominantly Buddhist faith, collectivization, undermining of cultural links with Mongolia and Russian immigration. Tyva has in fact been one of the more confrontational republics in its relations with Moscow. Its 1993 constitution claims that the republic is sovereign, retains the right of secession from the Russian Federation and bases its relations with the Federation on a treaty basis. All of this is contrary to the Russian constitution. However, rather than seeking full independence, this underdeveloped republic may find it within its interests to continue as part of Russia, trying to secure more resources for development while maintaining its cultural identity.

Buryat Republic The rise of Buryat nationalism in the early twentieth century and its pan-Mongol tendencies were commented on above. The Buryats today are easily outnumbered by the Russians in the republic (but not in Aga-Buryatia). Buryatia itself has a fairly diverse economy, based on agriculture, mining, forestry, engineering and other activities. The recent revival of Buryat national feeling has focused on a restoration of the pre-1937 frontiers for the republic, encouragement for Buddhism and other emblems of Buryat culture, environmental conservation (especially of Lake Baykal) and the deepening of relations with the Mongols.

Sakha Republic (formerly the Yakut ASSR) The huge Sakha Republic in the Far East economic region occupies some 3.1 million square kilometres, or over 18 per cent of the Federation's territory (figure 9.5). Its total population in 1995 was 1.035 million, a decline of 58,000 since 1992. The Sakha are a Turkic people who originally migrated from the southwest with their horses and cattle. Retaining pastoral nomadism, they adopted reindeer herding from the northern peoples and practised it in those regions where the environment was too harsh for their traditional practices to flourish. The Russians began to move into the region in the seventeenth century, seeking furs and other resources, but even as late as 1926 the Sakha formed over 80 per cent of the population of the republic, and the Russians only 10 per cent. Sakha nationalism began to assert itself in the early twentieth century among some local intellectuals, and the Yakut ASSR was established in 1922. In the Soviet period, Russian and associated immigration grew rapidly as part of the industrialization drive. Gold, coal, timber, iron ore and other resources proved important attractions and, from the 1950s, diamonds. By 1989, the Sakha formed only 33 per cent of the republic's population and the Russians 50 per cent. The northern part of the republic also contained about 25,000 of the Peoples of the North, namely Evenki, Eveny and Yukagiry.

Today, Sakha's economy focuses on resource extraction, particularly non-ferrous metals and minerals. In the early 1990s the republic was producing almost all Russia's diamonds, about a third of its gold, much

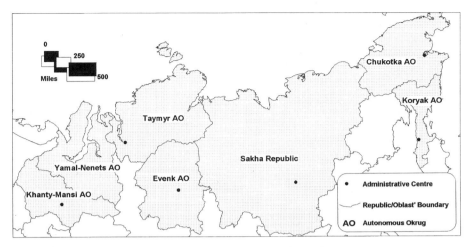

Figure 9.5 Autonomous territories of northern Siberia and the northern Far East.

of its tin and antimony, and coking coal, gas, timber and furs. Relations with Moscow have naturally focused around the control of this resource wealth, especially as it was argued locally that the republic had gained little from its years of exploitation by the Soviets. A revival of Sakha national feeling in the late 1980s was precursor to a declaration of sovereignty in 1990 and the adoption of a new constitution two years later. The recently elected president of Sakha agreed to sign the 1992 Federation Treaty, and according to the republic's constitution relations with the Federation are on a treaty basis. This, however, contradicts the 1993 Russian constitution. In 1995, the republic signed a bilateral agreement with Moscow, which included provisions for revenue sharing from diamond and other sales. Clearly, this resource-rich republic is in a position to gain considerably from taking a tough negotiating stance with Moscow, but its geographical position probably restrains it from seeking full independence.

Some of the northern minority groups within Sakha have likewise been arguing for more autonomy, but this time from Yakutsk. Some have been seeking a restoration of their national okrugs, which were established in the 1920s but abolished by Stalin in the late 1930s. Interestingly enough, many of these northern peoples have adopted Sakha as their first language, rather than Russian.

Territories of the Far North

As noted in chapter 3, the minority peoples of the Russian north are of a varied ethnic background. There are, however, numerous similarities in their economies and way of life, and many continue to live by such traditional pursuits as hunting, fishing and reindeer herding. The 1989 census listed 26 nationalities, though more than 30 are known to exist (Fondahl, 1995). The north contains seven autonomous okrugs and only seven northern peoples are represented by them. Thus the majority of nationality groups lack an official homeland, living in one or more of the homelands of their northern neighbours, or in the Sakha Republic, or in a non-ethnic region like Murmansk or Magadan oblasts. Needless to say, such groups have particular difficulty maintaining their numbers and identity. Problems also arise out of the geographical dispersal of some of the northern peoples. The Evenki, for example, numbered some 30,000 in 1989, but only 12 per cent lived in their autonomous okrug, where they formed 14 per cent of the total population. The others were scattered across adjacent territories – many more lived in Sakha Republic, for example, than in the official homeland (figure 9.5).

Tsarist policy towards the northern minorities was generally paternalistic. The Speranskiy code of 1822 accorded them special protection, but this proved difficult to enforce in practice. Soviet policy varied from the paternalistic to the downright predatory. Thus, in the early days of Soviet power, there were those who argued for continued protection of the northern peoples and preservation of their way of life, while others felt that their best hope lay in their absorption as soon as possible into the mainstream of Soviet life. From 1924 to 1935, the northern peoples were overseen by an official Committee of the North, whose remit was to introduce social and educational welfare. Gradually, however, this enlightened approach was overtaken by the industrialization drive and the determination to find and exploit the region's natural resources. The northern peoples particularly suffered from collectivization, which interfered with their traditional activities and communal existence, especially when accompanied by attempts at denomadization. Even more disruptive were the campaigns for resettlement and denomadization in the 1950s and 1960s, when Khrushchev's policy of reconstructing the rural settlement pattern (see chapter 7) was extended to the north. The result was the splitting up of northern families and the discouraging of youngsters from following in the footsteps of their parents (Vitebsky, 1996). Meanwhile, the northern peoples found themselves swamped in their own homelands by the tide of white immigrants, a process which continued right down to the late 1980s. In 1989, the percentage of Russians in the populations of the seven northern okrugs varied from 59 per cent in the Yamalo-Nenets Auronomous Okrug to 67 per cent in the Taymyr and Evenki okrugs. Nowhere did the titular nationality exceed 16 per cent of the population (table 9.2).

During most of the Soviet period the northern peoples were isolated from the outside world, few foreigners were allowed to venture north and even Russian movement was controlled. Little was therefore known about the effects of Soviet policy, other than official propaganda, which was uniformly positive. In the late 1980s, however, a more comprehensive picture began to emerge. It became apparent that resource exploitation had brought few tangible benefits to the northern peoples, who often found it difficult to adapt to urban and industrial life and thus generally did not move to take up the new forms of employment. The Soviet mode of development, therefore, effectively meant the establishment of predominantly Russian urban–industrial settlement nodes in a rural sea where the native peoples tried to continue with their traditional ways. This, however, proved increasingly difficult to do, especially in view of the severe environmental impact of many northern developments (see chapter 6). The northern environment, as we have seen, is particularly delicate, and the economies of northern peoples are acutely dependent on it. Processes such as the burning and pollution of reindeer pasture through industrial accidents, interrupting reindeer migration routes by building pipelines, polluting water bodies and forest removal can therefore have particularly devastating consequences for northern peoples. The health and well-being of the northern minorities also gave cause for concern. Low birth rates, high death rates, high levels of morbidity, alcoholism, family breakdown and crime were among the list of problems that were detected. In addition to the environmental and economic factors mentioned above, radioactive pollution, the demands of the military and inadequate investment in health care and social welfare were all implicated. Some northern peoples actually declined in absolute numbers between the 1979 and 1989 census dates, and assimilation was not the only reason. Where assimilation did occur, however, an accusing finger was increasingly pointed at deliberate Russification and the neglect of native cultures.

As is the case with other ethnic groups, the period since 1985 has seen a growing assertiveness by the northern peoples, demanding recognition of their problems and grievances as well as a greater share of the national cake. An Association of Peoples of the North, established in 1989, began to speak out for the northern peoples as a whole, and a host of local organizations began to do the same for individual regions or ethnic groups. Growing international contacts also led to a higher profile in international and foreign forums and to aid and advice from groups who had experienced analogous problems in other countries. Northern peoples (and also local Russian elites) began to lobby for greater autonomy, not only in relation to Moscow but also in relation to the local administrative centres to which the autonomous okrugs are subordinate. Thus Chukotka gained its independence from Magadan Oblast in 1992, and the other okrugs have been arguing for a similar change in status. Peoples without official homelands have been pushing for them. Thus Eveno-Bytantayskiy Rayon in Sakha Republic, abolished in the 1930s, was re-established in 1989.

Economic policies pursued by the Russian government since 1991 have brought the north into crisis (see chapter 5). The closure or reduction of many resource-based activities, rising costs, falling living standards, increasing isolation, defence cuts and other problems have encouraged a flight to the south, especially by Russians. Many northern regions face acute supply problems each winter. It may be that some of this is to the advantage of the northern peoples, since it represents a reversal of the development trends of the Soviet period. Equally, however, unless they are to become completely isolated and revert to entirely traditional patterns of life, the northern peoples do need forms of economic development and regional assistance. The question must be whether the Russian government can afford to help them and whether they themselves are in a position to benefit from any future economic developments which may occur. To date, Russia's northern minorities have a considerable way to go before they can hope to benefit from the proceeds of resource exploitation and other activities like their peers in Alaska and northern Canada. Much remains to be decided about property rights, local governance, environmental protection and other issues (Fondahl, 1995; Osherenko, 1995). Self-government by the minorities themselves is still in its infancy. Whether Russia's new market economy will benefit the northern peoples (and at what price), whether it will pass them by and whether the northerners will succeed in sustaining their cultures and traditional ways of life are all questions whose resolution will profoundly affect the north in the twenty-first century.

Peoples without Homelands

As noted above and in chapter 3, many nationalities in Russia have no official homelands. Such peoples include minorities from other former Soviet states who migrated to Russia in the Soviet period or even earlier: Ukrainians, Belarusians, Armenians and others. Many of these people have recently been returning to their own states. Other groups without homelands are considered too small or dispersed to warrant them. Three groups require brief comment.

The Germans In 1989, just over two million ethnic Germans resided in the USSR, 842,000 of them in Russia. The Germans are descendants of voluntary migrants who moved by invitation from their German homelands in the second half of the eighteenth century and the first half of the nineteenth to take up farming on the steppe, or sometimes urban trades. In Russia they were settled in special colonies and maintained their own communities. In the 1920s, a special Volga German ASSR was set up along the lower Volga for the large German community in that area, but this was abolished in 1941 in the wake of the Nazi German invasion of the USSR. The Volga Germans and other Soviet Germans were exiled to the east, mainly to Siberia and Kazakhstan. Only in the 1960s were they rehabilitated, but most remained in their new locations. Migrations to Germany began in the 1970s and again became significant in the late 1980s. From the late 1980s there have been calls for restoration of the Volga German republic and other territorial rights, but failure to secure these (partly because the former German settlements in European Russia have long since been occupied by others), coupled with the prevailing economic climate, have done little to dissuade Germans from emigrating.

The Jews There were 1.4 million Jews in the Soviet Union in 1989, a drop from 2.2 million in 1959. Russia's total in 1989 was 537,000. Most of the country's Jewish population was originally acquired in the eighteenth-century partitions of Poland, and during tsarist times most of them were compelled to remain in the empire's western provinces. In the Soviet period they became more dispersed and many more moved into cities. The unsuccessful attempt to establish a homeland for them in the Far East was discussed above. They suffered grievously at the hands of the Nazis in the 1941–5 war. Jewish numbers began to drop in the 1970s as a result of emigration. Mass emigration to Israel, the USA and other countries has continued in the recent period as a result of the economic situation and fears of anti-Semitism.

The Cossacks The cossacks are not a separate ethnic group from the Russians or Ukrainians, but are descended from pioneers of the steppe who fled from the dominions of the tsar and the king of Poland and adopted the lifestyle and military methods of the Tatars and other steppe nomads. Eventually they assumed the role of armed guardians of the frontier, with military formations acting as auxiliaries to the regular

Russian army. They had their own communities in various frontier locations. Suppressed by the Soviets, they have recently been reasserting themselves and insisting on their right to resume their historic semi-military role and their former territorial status. In June 1992, President Yeltsin signed a decree rehabilitating the cossacks and permitting the establishment of cossack units in the army and frontier guards. The cossacks are divided into different formations: Kuban' cossacks, Terek cossacks, Don cossacks and others.

Conclusion

Like Russia's non-ethnic regions, the autonomous territories are engaged in a political power struggle with the centre. The issues include the normal regional problems of political autonomy, ownership of resources and property, taxation and the right to benefit from the new market economy. But unlike the non-ethnic units, the autonomous territories are also concerned with ethnic and cultural matters: the survival of their distinctive cultures and lifestyles, relationships with non-titular residents, language, the right of an ethnic group to determine its own future. The issues involved are inevitably complex and, as this chapter has tried to show, are different for each republic and autonomous region, not to speak of the ethnic groups without homelands. On the one hand, each republic and autonomous region must ponder how much freedom it wants from Moscow, which means considering how viable it would be as a separate state given the issues of non-titular residents, geographical location, the need for access to the outside world, for defence, for outside assistance and so on. Equally, Moscow must consider its own interests: the territorial integrity of the country, its need for resources, issues of access and the costs of enforcing its rule in different regions. As the case of Chechnya has shown, those costs can be considerable. Hence, unless various parts of the country are to break away, compromise is called for, and perhaps different compromises are appropriate for different territories. This is in fact what the 'asymmetrical federalism' which lies at the heart of the present system (see chapter 3) is all about. Whether this will be viable in the long term remains to be seen.

The Bolsheviks originally hoped for an international society in which ethnicity would mean little or nothing. In the event, by establishing ethnic regions, they entrenched ethnicity into the Soviet system, and it now seems too late to weaken it. Even if some parts of the country break away in the future, Russia will remain a multinational, multicultural state. It can take some comfort from the fact that it exists in a multicultural and increasingly pluralistic world. Whatever the proponents of the 'globalization of culture' may assert, it seems very unlikely that either Russia or other countries will be able to rest content with simplistic assumptions about cultural convergence, at least for the foreseeable future.

Ten

Russia and the 'Near Abroad'

> The time has come for an uncompromising *choice* between an empire of which we ourselves are the primary victims, and the spiritual and physical salvation of our own people.
>
> *Alexander Solzhenitsyn, 1991*

Chapter 9 suggested that the republics of the Russian Federation are involved in a power struggle with Moscow. Some of them, it might be argued, are now engaged in a process of 'nation building' – the attempt to construct a separate identity based upon a sense of common history and community, and a common destiny for the future (Smith, 1991). It remains to be seen whether this process will yet lead to the break up of the Russian Federation.

Similar processes of nation building have been going on beyond the boundaries of the Russian Federation in the 14 other post-Soviet states (the former Union republics). This is an area to which Russians frequently refer as the 'Near Abroad', since these states have only recently joined the category 'abroad' and most Russians still feel them to be close geographically, culturally and historically. The difficulties which many Russians have had in facing up to a post-imperial future, with its implication of a diminished role on the world stage, have been outlined in chapter 3. The temptation for Russians to interfere in the internal affairs of their neighbours and to try to draw the post-Soviet states into what political geographers call Russia's 'sphere of influence' has been particularly great. It is a temptation which other post-imperial states have also felt and have responded to by creating such institutions as the British Commonwealth and the French Union. For their part, most of the post-Soviet states have been trying to make their newly won independence a reality, or at least to enter into closer relations with their gigantic neighbour only on their own terms. Relations between Russia and the other post-Soviet states are therefore particularly delicate, and are likely to have considerable influence on Russia's evolution and developing role in the world. This is the substance of the present chapter.

Nation Building in the 'Near Abroad'

Earlier chapters have suggested that Russian imperialism, while very much falling within the tradition of European imperialism which began in early modern times, had some peculiarities of its own. For geographical reasons, Russia's empire was never a simple set of relationships between a 'mother country' and its colonies, since it was never entirely clear where the 'mother country' ended and the colonies began. Some scholars have even claimed that Russians before 1917 had relatively little sense of ethnic prejudice or cultural superiority relative to the minorities within the empire. This is a sweeping claim with little validity as a generalization, although no doubt it is more true in some cases and periods than in others. Some scholars have also argued that, unlike what might be expected from the classical model of economic imperialism, some regions inhabited primarily by Russians actually suffered economically (especially during the Soviet period), to the benefit of parts of the periphery. This view is reflected in Solzhenitsyn's statement, quoted above. It stands in a tradition of Russian liberal thought which has long claimed that Russians and non-Russians, before and after 1917, have alike been victims of the Russian state. Be that as it may, the fact remains that in both periods the state was dominated by Russians and reflected the culture and aspirations of the Russians more than of any other people. The foregoing points serve to remind us that imperialism is an immensely complex phenomenon which can rarely be reduced to a simple, one-way set of relationships. It is, nevertheless, imperialism.

It is therefore hardly surprising that the post-Soviet states should tend to regard themselves as post-colonial societies needing to adopt nation building policies. In many cases the beginnings of these policies predate the 1991 break-up of the USSR. Thus, even while Gorbachev was still in power, most republics began to reconstruct their governments, establish presidencies and adopt such symbols of statehood as new flags and non-Russian place names. Since 1991, such tendencies have continued, with states developing new laws for citizenship, giving precedence to the titular language, reorganizing their educational systems and cultural policy so as to emphasize a new national outlook and distinctive national achievements in the arts and sciences, attempting to enter the global economy and seeking new international relationships to replace the unilateral dependence on Moscow which characterized the Soviet era.

Measures such as these are in line with the theory of post-colonialism, which has attracted the interest of human geographers in recent years (Pratt, 1992; Godlewska and Smith, 1994). The theory analyses the ways in which post-colonial societies seek to redefine their identities and escape the intellectual, political and economic dependency of their colonial pasts. Rewriting the histories of post-Soviet peoples so as to challenge long-prevailing Soviet and Russocentric interpretations of the past, and seeking other ways of asserting the cultural validity and distinctiveness of the colonized societies, are processes taking place in all the post-Soviet states, as in other parts of the world which have experienced colonialism.

As is almost always the case with the transition from colonialism, there are a number of factors which may hinder the decolonization, and these factors may be more significant for some of the post-Soviet states than for others. Those states, for example, which have recent experience of political independence or very distinctive cultures and histories may find things easier than others. The presence of a significant Russian population will certainly be a problem. We may recall from chapter 1 that the territories of most of the non-Russian states correspond with the former 'nationalistic empire', where initial Russian settlement was minimal. But where such territories overlap with the 'settler empire' (as in southern Ukraine or northern Kazakhstan) or where there has been much Russian settlement in the Soviet era (as in Estonia and Latvia), then there is a Russian minority problem. Other factors which may influence decolonization include the extent to which a new state's economy is viable (the presence of natural resources is a big plus), the degree to which the state has friends and allies abroad and geographical and security issues.

The three Baltic states of *Estonia*, *Latvia* and *Lithuania* (figure 10.1) have probably found it easiest to assert their independence from Russia, even though the first two have significant Russian minorities. The three states are non-Slavic and non-Orthodox and have close historical associations with Scandinavia, Germany and Poland. Although falling within the Russian sphere in the eighteenth century, they managed to maintain their distinctive cultures and ways of life, and there was a rise of national feeling in the nineteenth century. Most significant, however, was the fact that the three achieved political independence between 1918–20 and 1940. The legality of their annexation by Stalin in 1940 was challenged in the international community, and a sizable expatriate population campaigned for their independence thereafter. The peoples of the three republics were also aware of their relative underdevelopment in the Soviet period by comparison with their north European neighbours, who had had comparable living standards before the war. Furthermore,

Figure 10.1   The Baltic states and Belarus'.

Estonians and Latvians worried about the effects of Russian immigration. These are some of the reasons why the three Baltic states were in
the van of the independence movements of the late 1980s and early
1990s (see chapter 3) and achieved their independence even before the
Soviet break-up, in September 1991.

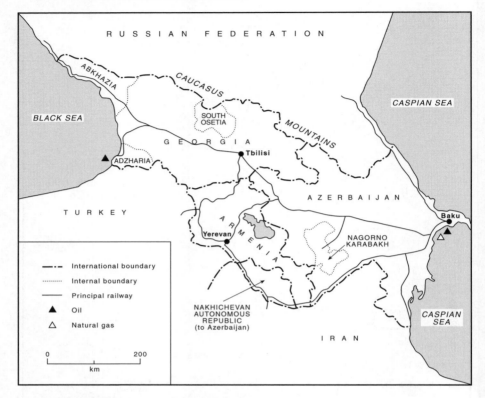

Figure 10.2   The Transcaucasian states.

The three Transcaucasian states of *Georgia*, *Armenia* and *Azerbaijan*
(figure 10.2) are also culturally and ethnically distinctive, both from
Russia and from one another, and although they mostly came under
Moscow's rule in the late eighteenth and early nineteenth centuries, all
three knew a short period of independence immediately after the Russian
Revolution. Historically, feelings towards Russia have tended to be
ambivalent: gratitude towards Russia on the part of Christian Georgia
and Armenia for preserving them from their Islamic neighbours Turkey
and Iran, and on the part of Islamic Azerbaijan for its modernizing role
(the oil town of Baku was a cosmopolitan community even before 1917),
but resentment at Russian imperialism and lack of cultural and political
sensitivity. During the Gorbachev period, the ethnic cauldron which is
the Transcaucasus and the North Caucasus boiled over, and relations
with Moscow were ruined as a result of the latter's insensitive attitude.
Independence therefore became inevitable. Ethnic issues will no doubt
continue to be influential in the future, but Georgia and Armenia may

Figure 10.3   Kazakhstan and Central Asia.

find themselves isolated without some kind of rapprochement with Russia.

The four Islamic republics of Central Asia – *Uzbekistan, Kyrgyzstan, Turkmenistan* and *Tajikistan* (figure 10.3) – are also culturally very distinctive, and only fell into the Russian sphere in the nineteenth century. They knew no national existence before the Soviet period, and even then family, clan and regional differences continued to be important. Although there was some growth of anti-Moscow feeling in the Gorbachev years, the evidence seems to show that the people of these states were not notably desirous of independence in 1991, being economically very dependent on the rest of the USSR. There have therefore had to be some painful readjustments since.

While also formally attached to Islamic Central Asia, Kazakhstan (figure 10.3) is more of a transitional state, in the sense that it contains a large Russian minority and was particularly closely tied in to the Soviet economy. The Kazakhs are a Turkic people with a distinctive culture, but for economic and demographic reasons Kazakhstan finds it more difficult than its southern neighbours to distance itself from Moscow.

The small republic of *Moldova* (formerly Moldavia) (figure 10.4) has

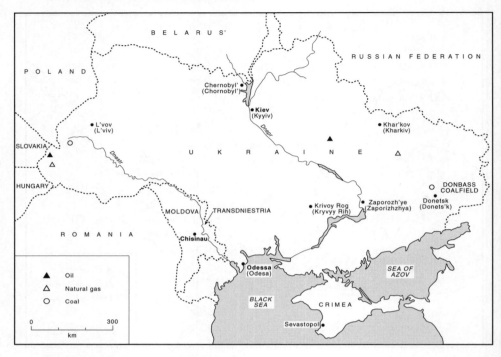

Figure 10.4 Ukraine and Moldova.

only known existence as a Union republic since 1940, when most of the territory on which it stands was annexed from Romania. Since Moldovans are ethnically Romanian, the republic had great difficulty in the Soviet period in justifying its existence. Since then, there have been calls for reunification with Romania, but more recently these have been tempered by the feeling that there are advantages in remaining separate. Moldova, for example, does not border on Russia and therefore may feel less threatened than some other states, and may continue to reap benefits from trade. The republic also has large minorities of Russians and Ukrainians who would certainly feel antagonized by reunion with Romania.

Chapter 3 made the point that Russian nationalists tend to regard *Ukraine* and *Belarus'* (figures 10.1, 10.4) as a natural part of the Russian realm, joined to it by many ties of ethnicity, history, culture and religion. Their declarations of independence from the USSR after the failed coup in August 1991 therefore came as a severe blow. In actual fact, the Ukrainians had experienced rising nationalism since the nineteenth century, particularly centred in western Ukraine, which has a distinctive religious culture (Uniat) and which only fell into the Soviet sphere after 1939. Most of Ukraine, however, was annexed to Russia in the seventeenth and eighteenth centuries, and there was considerable Russification of the eastern and southern parts in particular. Like the Transcaucasian states, Ukraine knew a brief period of independence after the Russian Revolution. Since 1991, the Ukrainians have tried hard to assert their distinctiveness and right to independence from Moscow in the face of the hostility of Russian nationalists and many communists. The independence of Belarus' is if anything made more difficult by a greater degree of Russification and by the recent actions of a left-leaning president who favours much closer ties with Russia.

The break-up of the USSR inevitably gave rise to many problems, and the establishment of the Commonwealth of Independent States (CIS) in December 1991 was meant to address them. One issue, for example, was the need to maintain a 'common economic space' in view of the close interdependence of the republics under the Soviet centrally planned economy. Disruptions to this 'common economic space' had already contributed to the economic turmoil of the later Gorbachev period and were to continue to do so after 1991. This matter is addressed in chapter 4. Another problem was the need to sort out claims to Soviet assets and also to distribute the former USSR's debts among the successor states. Yet another was the military and security issues inherited from the USSR. The rest of this chapter will consider three long-term issues which have influenced relations between Russia and its post-Soviet neighbours since 1991, and then return to the question of the future of the CIS. The three issues are: the Russian diaspora; territorial and security problems; and economic and trade relations.

The Russian Diaspora

The break-up of the USSR in 1989 left some 25 million ethnic Russians living in the 'Near Abroad' (Kolstoe, 1995; Kaiser and Chinn, 1996). The distribution of ethnic Russians in the other post-Soviet states, and their numerical significance in those states, is described in table 10.1. Since that time many Russians have migrated to the Russian Federation. By the middle of 1996, it was estimated that the net migration to Russia from the other post-Soviet states since 1989 exceeded three million people (by no means all ethnic Russians) (Dmitriev, 1996; Hilton, 1996). Of these, however, only just over one million were officially registered.

The fact of ethnic Russians living in the 'Near Abroad' must inevitably influence Russian policy towards those states. For one thing, there

Table 10.1   The Russian diaspora in the post-Soviet states, 1989

Republic	No. of Russians (thousands)	Russians as % of total population of republic
Ukraine	11,356	22.1
Belarus'	1,342	13.2
Estonia	475	30.3
Latvia	906	34.0
Lithuania	344	9.4
Moldova	562	13.0
Georgia	341	6.3
Armenia	51.6	1.6
Azerbaijan	392	5.6
Kazakhstan	6,228	37.8
Uzbekistan	1,653	8.3
Kyrgyzstan	917	21.5
Turkmenistan	334	9.5
Tajikistan	388	7.6

Source: 1989 Census of Nationality.

is the normal desire to protect one's ethnic cousins, a desire which tends to be felt most strongly by nationalists. People who resent the break-up of the USSR and the consequent diminution of Russian power may feel tempted to interfere in the internal affairs of other states, using the existence of minorities as an excuse. Similar temptations may be felt by those who are concerned by defence and security issues. Equally, however, Russia, must be worried about the implications of further mass migrations to Russia, since these have already caused enormous problems in housing provision and employment (see chapter 7). In November 1992, a presidential decree accorded protection to ethnic Russians living in the 'Near Abroad'. In reality, however, there is probably relatively little that can be done to dissuade those who feel the need to move to Russia from doing so. The Russian law on citizenship of 1992 grants citizenship to all who were born in the Russian Federation and to those born elsewhere in the former USSR who are not already citizens of other states. The 1993 laws on refugees (those not qualifying as Russian citizens) and on forced migrants (those qualifying) enjoins hospitality to be shown to these groups (Russia signed the UN Convention on Refugees early in 1993). Whether these provisions will be fully realized in practice, however, is another matter, given shortages of funds. Recent policy has favoured trying to persuade Russians to remain in the states where they now live. Even so, some forecasts suggest a potential net migration to Russia of between two and five million by the year 2,000.

The largest Russian minority is in *Ukraine,* where 11.4 million were living in 1989, forming 22 per cent of that republic's total population.

As noted in chapter 1, the origins of Russian settlement in Ukraine lie in the eighteenth and early nineteenth centuries, when both Ukrainians and Russians had a role, together with other nationalities, in the settlement of the southern steppe. Later, during Russia's industrial revolution, many Russians migrated to the new industries being established in the Donbass and in other parts of southern Ukraine. By the time of the first Soviet census in 1926, Russians constituted 9.2 per cent of the population of Ukraine. Thereafter, with the exception of the war years, Russian immigration continued in the wake of Stalin's industrialization drive and subsequent developments. By 1989 they formed substantial minorities in Ukraine's eastern and southern regions, with the exception of Crimea, where they formed an absolute majority. Crimea, together with the eastern oblasts of Luhans'k and Donets'k, had a majority of Russian speakers, while other eastern and southern regions had substantial minorities.

A feature of Ukraine's political development since 1991 has been a split between eastern and western parts of the country, with western parts forming breeding grounds for ethnic Ukrainian nationalism and eastern parts favouring closer relations with Russia. Eastern Ukraine has often been suspicious of policies of Ukrainianization, such as those designed to foster the Ukrainian language. Some observers have therefore seen evidence of a potential schism in Ukraine, bolstered by calls for the federalization of the country and the granting of greater autonomy to the east. Such calls have found favour among those fearful of the effects of economic reform on the coal mining and heavy industries of the east and south. However, Ukrainian leaders have thus far been careful to pursue territorial forms of nationalism which would include ethnic Russians and Russian speakers. Thus citizenship was extended in October 1991 to all the country's residents, irrespective of their ethnic origin, and legal guarantees have been given to ethnic minorities. It would be unwise to assume that those registered as ethnic Russians in the 1989 census will continue to consider themselves Russian in the future. Evidence suggests that many residents of eastern and southern Ukraine are happy to identify with that country rather than with Russia. Such feelings may have been strengthened by a rather stronger Ukrainian economy, as well as by reactions to Russia's invasion of Chechnya. As long as overly narrow interpretations of Ukrainian national identity are avoided in the future, the country seems unlikely to split along ethnic lines.

Crimea is a special case. This region has a majority Russian population and is overwhelmingly Russian speaking. Moreover, it was only transferred to Ukraine in 1954, under rather dubious circumstances, before which it formed part of the Russian Federation. Russian and Ukrainian settlement of Crimea began after its annexation in 1783, but the area was so remote that it long proved difficult to settle intensively. Eventually, however, with the building of the railway, the exploitation of its unique environmental advantages for agriculture and beach tourism, and the development of the naval base at Sevastopol', the peninsula attracted a substantial population of immigrants. The fact that this strategic region was the historic homeland of the Crimean Tatars, however, proved a longstanding embarrassment, and they were eventually expelled by Stalin in the wake of the German occupation in the Second World War. Crimea's autonomy within Ukraine was promoted by the region's conservative, ex-communist and predominantly Russian elite. Autonomous republic status was granted in 1991, and eventually a bid was made for independence and closer ties with Russia. These moves found ready support among nationalists in the Russian parliament who pointed to the region's historic links with Russia and to its heroic role in the Crimean War and the Second World War. Meanwhile, however, political relations in the region were complicated by the now returning Crimean Tatars, who began to demand the restoration of their lands and of their former political status. The Crimean independence bid was effectively quashed by Ukrainian President Kuchma in April 1995, when he took control of the republic's government. Negotiations over the region's future status continue. While Russian nationalists both on the peninsula itself and in Russia may continue to hanker after reunion with Russia, harsh economic and political realities may dictate a less uncompromising stance on the part of the Crimean leadership. The Yeltsin government seems unwilling to take the side of the nationalists on this issue, though it may prove to be a useful tool for forcing concessions on Sevastopol' and related security issues in the future.

The second biggest Russian minority in 1989 was in *Kazakhstan*, where 6.2 million ethnic Russians were living, constituting 38 per cent of that republic's population. Russian settlement is particularly concentrated in the northern steppe oblasts close to the Russian frontier, and along the fringes of the mountains of the east. The capital Almaty (formerly Alma Ata) in the southeast was 59 per cent Russian in 1989 (56 per cent in 1993). Russians outnumbered Kazakhs in eight oblasts and formed an absolute majority in four (Harris, 1993a). As in Ukraine, Russian settlement in Kazakhstan has a long history. The steppe regions of the north long formed a borderland between the Russian agricultural settlement of the southern part of the West Siberian plain and the nomadic Kazakhs of the steppe and semi-desert to the south. As Russian settlement slowly crept southwards, the region was gradually transformed into part of Russia's 'settler empire' and the nomads were displaced. Russian agricultural settlement especially increased after the

Emancipation in 1861, and more especially after the building of the Trans-Siberian Railway in the 1890s and the Stolypin agricultural reforms in the early years of the twentieth century. Kazakh resentment at the loss of their historic grazing lands grew in consequence. By 1926, Russians already formed 19.7 per cent of the republic's population. Further Russian immigration followed in the wake of the Soviet industrialization drive, collectivization and the Virgin Lands agricultural colonisation scheme in the 1950s. Russians actually outnumbered Kazakhs in Kazakhstan by 1959, but this situation was reversed by 1989. This was largely because of higher birth rates among the latter, but also in part because of Russian net outmigration.

Russians in Kazakhstan tend to live in cities, except for the north and east, where they also form a substantial element in the rural population as a result of the settlement policies mentioned above. The latter regions also had many Germans, Ukrainians, Tatars and other non-Kazakhs in their populations in 1989. In general, the Russians and other Europeans do not intermarry with the Kazakhs, and the social lives of the two groups are frequently separate. The situation is thus rather different from that in Ukraine. Relatively few Russians speak Kazakh. Since independence in 1991, the Kazakh government has officially followed a non-discriminatory policy, conferring citizenship on all residents of the republic and protecting minorities. However, Russians have often felt intimidated by the nationalistic attitudes of some Kazakh politicians, while laws promoting the Kazakh language and the development of a new Kazakh indentity have often seemed threatening (Kaiser and Chinn, 1995). Russians and other minorities, traditionally privileged, have been losing out politically, in access to higher education and jobs and in other areas. There has thus been a drift of migrants back to Russia, more than a quarter of a million (net) leaving between 1989 and 1993. Kazakhs returning from other post-Soviet republics and also from China (where many fled after 1916–17) have been given citizenship and settled largely in Russian-dominated areas. Similarly, the policy of moving the capital from Almaty to Akmola (Aqmola, formerly Tselinograd) in the north has provoked fears that this is an attempt to dilute the Russian population. As an alternative to leaving, some Russians in the north have argued for autonomy for their regions or even for a frontier adjustment in favour of Russia. Needless to say, the latter idea is supported by numerous nationalists in Russia. Thus the potential exists for friction between Russia and Kazakhstan over this issue. Since the majority of Russians in Kazakhstan have lived there from birth, they are reluctant to leave. Whether Kazakh nationalism can be broad enough to embrace them remains to be seen.

Kyrgyzstan in Central Asia also has a sizable Russian community, forming 21.5 per cent of the republic's population in 1989 (917,000 people). Most live in and near the capital, Bishkek, but they also form a notable proportion of the rural population in the east. Their problems have been rather similar to those of Russians living in Kazakhstan, and there has been a substantial outmigration. Recent policy has been directed at persuading them to stay because of a shortage of skilled labour. Thus the Russian language was accorded a special status in 1994 (which is also the case in Kazakhstan), and other measures have been taken to protect Russian culture. But these run counter to the feelings of Kyrgyz nationalists.

In the other three *Central Asian states*, Russians have formed even more of a minority – all three republics had fewer than 10 per cent Russians in 1989. With their predominantly urban lifestyle, jobs mainly in industry and white collar professions, ignorance of local languages and numerous privileges, the Russians of Central Asia seemed even more of a colonial elite than was the case in other republics. Their loss of status has come as a shock to which they find it hard to adjust, and there are fears of the consequences of nationalism, the Islamic revival and civil disorder (especially in Tajikistan). There has thus been considerable outmigration back to Russia. Policies to encourage Russians to stay, such as laws on dual citizenship passed by Turkmenistan and Tajikistan, seem unlikely to be very effective.

Russians formed only small and diminishing minorities in the three *Transcaucasian* republics in 1989. Many have now left as a result of the ethnic upheavals and other problems which have accompanied the fall of communism. In Belarus', by contrast, there were 1.3 million Russians in 1989, constituting just over 13 per cent of the total population. Their future within *Belarus'* seems secure at the moment, with a quite Russified population (20 per cent of the ethnic Belarusians in 1989 spoke Russian as their first language) and the activities of a pro-Russian president.

One of the most complex situations to face ethnic Russians has been in the three *Baltic states*, especially in Estonia and Latvia. All three had been part of the Russian Empire before 1917, and both Estonia and Latvia had sizable Russian minorities in the 1930s (forming 8.2 and 10.6 per cent of their populations respectively). After their annexation by the USSR in 1940, they all received considerable inflows of Russians, while some of their own peoples were exiled to other parts of the Soviet Union. Russians were needed to work in the new heavy industries of Estonia and Latvia, where low rates of natural increase in the population had given rise to labour scarcity. Lithuania, with its predominantly Catholic population, and therefore higher birth rate, and its later industrialization, experienced lower levels of immigration. Russians were also attracted to all three republics by their higher living standards and European lifestyles. By 1989, ethnic Russians constituted over 30 per cent of Estonia's population, 34 per cent of Latvia's and 9.4 per cent of Lithuania's, a total of 1.7 million people in the three republics. The sheer scale of Russian immigration into Estonia and Latvia gave rise to real concerns about the future of their national languages and cultures, and since 1991 over whether national independence could be made secure. Both Estonia and Latvia thus adopted ethnic rather than inclusive concepts of citizenship. Denouncing the 1940 annexations as illegal, both states granted citizenship only to the descendants of those who had lived there before that date. Others had to fulfil both residence and language requirements. These provisions obviously discriminated against the Russians, relatively few of whom were descended from those who had lived there before 1940 or spoke the national languages. Russians found themselves barred from participating in national politics, and there have been problems in such areas as securing employment, owning property and travelling abroad. Russians have shown their usual resentment against policies to promote national identity, although both internal and external pressures have helped to secure important rights and concessions in both states. In Lithuania, where an inclusive definition of citizenship has been adopted, the situation is somewhat better, but all three republics have been experiencing some Russian outmigration. The fact that in Estonia and Latvia there are sizable concentrations of Russians close to the eastern frontier with the Russian Federation has led to calls by Russian nationalists for frontier revision. The predominantly Russian speaking city of Narva in northeast Estonia held a referendum on autonomy in July 1993. The Russian government has used the ethnic issue to bring pressure to bear on the Baltic states. However, there is evidence that many Russians are now beginning to adjust to the new situation and may wish to identify with the states where they live. Certainly, the prospects of higher living standards as the Baltic states marketize and seek membership of the European Union may help to counteract disaffection.

Moldova's Russian population in 1989 was 562,000, or 13 per cent of the population. The republic also contained rather more Ukrainians, as well as several other minorities. The major problem here has been the attitude of that portion of the Russian minority (about a third) which lives on the left bank of the Dnestr River, where, together with many Russified Ukrainians and Moldovans, a definite Slavic culture is to be found. The population of this territory resented policies for promoting a Moldovan national identity and plans for reunification with Romania. The proclamation of a *Transdniester Republic* in September 1990 eventually led to open conflict, with Russian nationalists and cossacks taking the side of the Transdniestrians. A somewhat dubious role was played by the Russian Fourteenth Army under General Alexander Lebed. An agreement between Russia and Moldova was signed in July 1992 to normalize the situation, and a basis for regional autonomy was agreed in April 1994. However, at the time of writing full agreement has not yet been reached, and the area is still occupied by a contingent of Russian troops.

Territorial and Security Issues

The issue of Russian minorities in the 'Near Abroad' is part of a broader series of territorial and security problems which the collapse of the USSR has brought in its wake. In the Soviet period, the internal borders between the Union republics had little significance and could be freely crossed. By contrast, the Soviet authorities expended enormous human and material resources in guarding its external frontiers, even those with 'fraternal' socialist states like Poland and Czechoslovakia. Nowadays there is the need to secure the borders of the Russian Federation, not only for military and security reasons, but also to control migration and trade, cut down on smuggling and crime and for other reasons. This is an enormous expense, but made rather easier by the fact that Russian border guards still patrol much of the old external frontier of the USSR under the auspices of the CIS. Considerations of security also mean that the Russian government must be concerned about what happens beyond its frontiers in neighbouring states. The Russian state frontier is long and difficult to patrol, and problems in neighbouring states can easily spill over into Russia, especially when culturally similar populations are found on both sides of the border. It is also impossible to ignore old-established traditions of power politics and desires for 'spheres of influence' at this point. Great powers have always been attracted by the idea of securing informal control over territories where they can achieve military and possibly also economic goals without the expense of a full occupation. Such aims were familiar to the architects of both Russian imperial and Soviet foreign policy and may appeal to their successors, although those successors might also be influenced by other views of what best constitutes Russia's national interests.

One of the most worrying security problems for post-Soviet Russia has been the series of ethnic conflicts and wars which have raged in the republics to its south since the late 1980s. One aspect of this has been the resurgence of Islam. After decades of communism, the peoples of the region have been turning to their traditional religions as markers of national identity. The movement can be seen as part of the worldwide revival of Islam which has been occurring since the 1970s. One worry for Moscow has been that it might lose influence in the southern states to such Islamic powers as Iran, and perhaps even find itself confronted along its southern boundary by a series of hostile states. The Islamic revival also has the potential to add fuel to the flames of nationalism and separatism within the Russian Federation itself. As noted in chapter 9, Russia contains many Muslim peoples, including most of those in the North Caucasian republics as well as peoples in the Middle Volga area such as the Tatars and the Bashkirs. Unrest in these regions would have serious repercussions on the Russian economy. It seems likely that it was fears of the demonstration effects of a successful secession from Russia by Islamic Chechnya which encouraged the Russian invasion of that republic in December 1994 (see chapter 9).

Russia has thus been busy attempting to reinforce its influence in the Islamic states, trying to mediate between Islamic Azerbaijan and Christian Armenia over the disputed territory of *Nagorno-Karabakh* and fostering closer ties with the Central Asian states through the CIS. Such policies have not only security dimensions but also economic ones: witness the 'mini-CIS' agreement signed with Kazakhstan, Kyrgyzstan and Belarus' in March 1996, or the bilateral agreement signed with Kazakhstan the following month. The latter will give Russia a major say in the development of Kazakhstan's Tengiz oilfield. Apart from Islam, one factor which has certainly encouraged Russian activity in Central Asia has been a fear that the region, with its largely Turkic heritage, might be drawn into the sphere of influence of a largely secular Turkey and through that of the West. Russian policy has therefore tended to bolster the power of the former communist elites who largely run these states, often by authoritarian means. The signing of military agreements with Turkmenistan and Kyrgyzstan has given Russia a role in building up their armed forces with the help of Russian manpower and supplies. While there is relatively little fear of Islamic radicalism in Kazakhstan or Kyrgyzstan, Russia's policy seeks to bolster the power of Central Asian leaders like Presidents Niyazov of Turkmenistan and Karimov of Uzbekistan, who follow cautiously pro-Islamic policies while distancing themselves from religious extremism. In Tajikistan, Russia along with Uzbekistan has supported the ex-communist government in its protracted civil war against democrats and Islamic fundamentalists. It has also been active in trying to protect the border with Afghanistan against incursions by guerrillas and other forces fighting in that country, which has a large Tajik population.

Russia is also worried by the ethnic upheavals in the Transcaucasus, which may similarly endanger its national security. Its mediation in Nagorno-Karabakh has been mentioned above, and its forces have also played a role in the disputes in South Osetia, Abkhazia and Mingrelia, all of them in Georgia (where Russia currently occupies four military bases). Because of the tortuous nature of these struggles, Russia has frequently fallen foul of local opinion, which suspects it of seeking to further its own interests in the region. In the case of the Caspian Sea, for example, Russia has been keen to assert its territorial rights over the exploitation of the sea's resources, which are also claimed by the bordering states of Azerbaijan, Kazakhstan, Turkmenistan and Iran. Policies to cultivate Russian influence in the Transcaucasus, which seem to stand their best chance in the Christian states of Georgia and Armenia (surrounded as they are by Islamic countries), are obviously complicated by these local disputes.

With one or two exceptions, Russia's relations with its western post-Soviet neighbours have been more peaceful, though frequently marred by mutual suspicions and tensions. The problem of Transdniestria in Moldova was discussed in the previous section. In the case of Ukraine, there has been much difficulty over the implications of having such a large Russian minority and the outspoken anti-Ukrainian sentiments of many Russian nationalists (including those in the Russian parliament) who would like to see the two countries reunited. Ukraine has been particularly suspicious of territorial claims (for example, those made by Russian nationalists to Crimea and parts of eastern Ukraine, though these are not supported by the Russian government). It sees these as manifestations of a potential resurgence of Russian imperialism. Thus Ukraine was initially reluctant to give up the nuclear arsenal it had inherited from the USSR, which complicated its relations with the United States and the international community. The problem of Crimea was discussed above. One significant aspect of this has been the issue of the naval base of Sevastopol', where the former Soviet Black Sea Fleet is based. Both Russia and Ukraine have ambitions as Black Sea powers and have had claims to the fleet. The division of the fleet, however, was complicated by the pro-Russian nationalism of many of the officers and crews as well as the inhabitants of Sevastopol'. Claims have been made, even at Russian parliamentary level, that Sevastopol' is rightfully Russian and was not transferred to Ukraine with the rest of Crimea in 1954. Perhaps not surprisingly in the circumstances, negotiations over the division of the fleet and the possible leasing of part of Sevastopol' by Russia dragged on for years. Relations between the two states have been made more difficult by these misunderstandings, as well as by the possibility that Ukraine might seek to enhance its security by joining NATO (traditionally an anti-Soviet, if not anti-Russian, alliance).

Relations between Russia and the three Baltic states have likewise been made difficult by mutual suspicions deriving from the history of the Soviet occupation, the Baltic states' pro-Western policies and refusal to join the CIS and the treatment of Russian minorities. Russia was initially reluctant to withdraw its forces from the region because of the last issue, and there have been numerous disputes over aspects of the Soviet heritage, such as property, pensions and environmental problems. There are a number of territorial disputes, including claims by Estonia and Latvia to territory handed over to the RSFSR after 1940, maritime borders and Russia's transit rights to the exclave of Kaliningrad (see figure 10.1). The possibility that the three states might join NATO similarly worries the Russians.

Economic and Trade Relations

The establishment of the Commonwealth of Independent States in December 1991 was by no means merely a sop to pro-Soviet sentiment. As we have seen, the collapse of the USSR left in its wake numerous problems which required attention. The original agreement provided for a coordinated foreign policy, a common economic space and cooperation in transport and communications, in environmental protection and in the fight against crime. There followed a collective security agreement, although this was signed by only six of the then eleven members (Georgia only joined the CIS in October 1993 and the three Baltic states refused to join, arguing that they should never have been annexed by the USSR in the first place). In the event, mutual suspicion militated against the effectiveness of several of the agreements. One aspect of this was the fear that Russia might use the CIS to establish its hegemony once again – the fact that Russia had over three-quarters of the territory, half the population and 60 per cent of the industrial production of the former USSR showed that these fears were far from groundless. There was also the determination by each of the successor states to pursue its own interests rather than see itself as a member of a collective.

The agreement on the common economic space was particularly vital in view of the way that the Soviet economy had been developed as an integrated whole, with a high degree of regional economic specialization, monopoly and hence regional interdependence (see chapter 4). Unfortunately, however, this was one of the agreements which failed to work particularly well, as communications broke down, trade barriers began to be erected and the common ruble zone disintegrated. The result was a sharp fall in inter-republican trade and a reorientation of Russian trade (newly liberalized along with the rest of the economy by the Yeltsin government) towards the outside world. Thus, whereas 70 per cent of Russian exports (by the then prevailing prices) went to the other Soviet republics in 1990, and 47 per cent of its imports came from there (the exports accounting for 12 per cent of Russian GDP in that year), only 4 per cent of GDP came from CIS exports by 1994. In 1996, exports to the CIS accounted for 20 per cent of all exports, and 30 per cent of imports.

Realignment of prices within the CIS towards world market levels has favoured Russia, with its large energy and other resources. Thus, whereas in the past inter-republican trade was dominated by industrial products like engineering and metal goods, electricity and fuels now dominate Russian exports to its neighbours. Rising fuel prices within the CIS have placed pressure on the other members who lack resources, and the result has been a growth in indebtedness. Astutely, Russia has been able to make use of this situation, acquiring key assets in other republics by way of compensation and using economic pressure to force conformity within the CIS and more generally. Russia does, of course, import food and other necessities from its neighbours, but it can also use the currency it earns from fuel exports to buy these things on the world market. In terms of the value of trade, Russia's biggest trade partners in the CIS are Ukraine, Kazakhstan, Belarus' and Uzbekistan. Of these, Ukraine accounted for 48 per cent of Russian exports within the CIS in 1995 and 43 per cent of its imports. Analogous figures for Belarus' were 22 and 20 per cent respectively.

The Future of the CIS

Since its foundation in December 1991, the CIS has been a rather weak and shadowy entity. CIS leaders and officials meet on a regular basis, and over 600 agreements have been signed since the community was born. But few of these are fulfilled in practice. Given the chaotic circumstances in which the CIS was founded, perhaps this is not surprising. On the one hand there were communists and Russian nationalists ready to denounce the collapse of the USSR and work towards its restoration in some form or other. On the other were those in all the republics who hoped for closer relations with various parts of the outside world and who feared a revival of Russian hegemony over the other republics. In March 1996, the Russian State Duma passed a resolution denouncing the 1991 Belovezha accords, which established the CIS and effectively destroyed the USSR. This convinced many outside Russia that a reassertion of Russian imperialism was under way, despite the fact that the resolution was dismissed by President Yeltsin as having no legal force. There can be no doubt that neo-imperialist ideas do still appeal to many Russians. Had the communist Gennadiy Zyuganov or the nationalist Alexander Lebed won the 1996 presidential election, it may well be that neo-imperialism would once again have been on the agenda. But, though Russia has sought UN approval for its self-appointed peace-keeping role within the CIS, there are many inside Russia who would caution against a neo-imperialist strategy.

Writing in 1994, Leslie Holmes described the CIS as 'particularly interesting, since it does not fit neatly into either domestic politics or international relations. Perhaps there is a need for a new intermediate category of politics for this – for relationships less close than in a federal system, yet closer than between genuinely sovereign states' (Holmes, 1995, p. 124). The point that Holmes was making was that the members of the CIS were still closely intertwined as they emerged from the ruins of the USSR. Even then, however, it was clear that strategies of integration within the CIS appealed to some of its members more than others. Thus Ukraine, ever suspicious of Russian intentions despite their mutual trading links, Uzbekistan, which regards itself as leader in Central Asia, Turkmenistan, with its energy wealth and official neutrality, Azerbaijan, with its oil, and Moldova, with its ethnic difficulties, tend to oppose further integration measures. By contrast, Kazakhstan, with its mixed population and close economic links with Russia, Kyrgyzstan, with its large Russian community, and Russified Belarus' are less negative in attitude. It was in fact with the latter three states that Russia signed a 'mini-CIS' agreement in March 1996. This envisaged closer economic and cultural integration between the four states. However, some sceptical observers link this agreement with the Yeltsin re-election campaign, since many CIS leaders were fearful of the prospect of the communist Zyuganov as president of Russia. It has been argued that the agreement could well undermine the existing rather nebulous links within the CIS, without actually bringing the four states much closer together.

A particularly striking example of integration within the CIS was the agreement to establish a Russian-Belarusian community signed in April 1996. The aim was to unify the two countries' economic and financial systems, stopping just short of an actual confederation. The moving spirits were President Yeltsin of Russia, who wished to use the agreement to undermine the integrationist stance of his major electoral opponent Gennadiy Zyuganov, and President Lukashenka of Belarus', who saw this as a way of propping up Belarus's ailing economy. While closer integration seemed to be supported by the Russian military and possibly also by its energy interests (with important oil and gas pipelines crossing

Belarusian territory), there were also concerns about the short-term costs to Russia and the difficulties of reconciling Russia's more marketized economy with the more conservative economic structures in Belarus'.

Since the foundation of the CIS there has been disagreement among its members over how far it is to be seen as a device for solving problems left by the collapse of the USSR, and how far a vehicle for promoting closer economic and political cooperation between its members. As is the case in the European Union, much depends on the politics of the individual as well as on the circumstances peculiar to each state. Unlike in the European Union, however, the waters have been muddied by the sheer chaos of the transition from communism. For Russia, in addition to the benefits to be gained from cooperation with its neighbours across a range of issues, arguments about military security and the restoration of Russian influence and control have had much appeal. Equally, however, voices have been raised warning of the costs to the Russian economy of signing hasty agreements with states which are economically more backward and conservative than it is. Such voices urge Russia to be more selective in its approach, perhaps seeking to attain its goals through bilateral agreements rather than through such sweeping arrangements as those connected with the CIS or the 'mini-CIS'. Clearly, this is an argument which is likely to run for some time. As integration between states seems to become a common response to the problems posed by a global economy, so forms of integration will obviously appeal to the USSR's successor states. But these may not necessarily take the form that they currently do within the CIS or the 'mini-CIS'. Indeed, without a much greater degree of economic success among CIS members, and especially Russia, prospects for their closer integration within the ambit of the CIS seem limited.

Eleven

Russia and the Wider World

> Everything that may plunge Russia into chaos and make her return to obscurity is favourable to our interests.
>
> *Louis XV of France, 1763, quoted in Anderson, 1987*

Marxism is a creed which calls for international socialism and has little regard for the nation state. We have seen, however, that its application to Russia in the circumstances of 1917, and particularly following Stalin's decision to build 'socialism in one country', led to that country's turning its back on the outside world. Even after 1945, when other communist states appeared, suspicion of the regions beyond the Soviet frontiers continued to reign. Not until the 1970s did the USSR begin to participate seriously in world trade, and not until the Gorbachev years was a concerted effort made to open up the economy and to dismantle other barriers which had long sheltered Soviet society from the world at large.

Since 1991, the process of opening up Russian society and economy to global influences has gone much further. Now more than ever it is important to view Russia in its international context. What is happening there, as well as in other parts of the world, is ever more closely linked to the global economy. As Manuel Castells wrote in the late 1980s, at a time when the Soviet Union still existed:

> What is new – is not that international trade is an important component of the economy (in this sense, we can speak of a world economy since the seventeenth century), but that the national economy now works as a unit at the world level in real time. – The coming integration into the world economy of Eastern Europe, the Soviet Union and China – will complete this process of globalization which, while not ignoring national boundaries, simply includes national characteristics as important features within a unified, global system. (Castells, 1996, p. 100)

The position which Russia occupies within that system will only partly be governed by global forces. Also important will be Russian perceptions of the outside world and the policies which Russia chooses to adopt towards it. These matters will be influenced by a host of factors, including how Russians view themselves and their place in the world now that the Soviet Union is no more. Before discussing such issues, however, it is important to say something about the foreign policy heritage which the USSR has left to the new Russia. The rest of the chapter will then consider the broader relationships which seem to be emerging between Russia and the outside world, focusing upon trade and security issues within the context of general geopolitical concerns.

The Soviet Heritage

Some scholars have argued for an essential continuity between the foreign policy pursued during the Soviet period and that likely to be adopted by post-Soviet Russia. This argument would be based on such factors as common geopolitical concerns (the fact that the Russian Federation occupies most of the territory of the former USSR), continuity in attitudes and institutions and even the fact that many of the people now helping to frame Russian policy also helped to frame Soviet policy before 1991. While accepting that there is some substance to this view, it is important not to overemphasize the continuities and to fall into the trap of determinism. Both Russia and the international framework in which it exists have changed since the late 1980s, and real choices face the Russian leadership. At the same time, it is clear that the Soviet period has left Russian policy-makers with many difficult problems in the international arena, problems which will have to be addressed one way or another.

Chapter 2 briefly described Russia's emergence as a superpower and major rival to the United States and its allies in the years after the Second World War. Ever since 1917, Soviet foreign policy had been governed by two principal considerations. The first was what we might call the traditional revolutionary principle – the desire to foster international revolution and support foreign communist movements whenever possible or deemed desirable. In the inter-war period this aim was furthered by means of the Communist International or Comintern, established in 1919. Later, other instruments were used, and even in the 1970s and 1980s Soviet support for various Third World revolutionary movements could be regarded as a continuation of this revolutionary tradition. The second consideration behind Soviet foreign policy, and one which seemed to become more important as time went on, was the pursuit of Soviet national interests. Among these interests, national security, the desire to demonstrate an international political and military equality with the United States and eventually the need to earn hard currency through trade loomed particularly large.

The USSR's great victory in the Second World War was not enough to ensure that it would feel secure. The Soviet leadership was still haunted by memories of how the western European states had refused to agree to mount a common challenge to Hitler's pretensions in the 1930s, and how in 1941–3 the Soviet Union had been left to face the Nazi onslaught virtually alone. Mutual suspicion between capitalism and communism, a suspicion which dated back to 1917, led to the splitting of the European continent between a capitalist west and a Soviet-dominated east. The countries which fell into the Soviet sphere were forced or otherwise persuaded to establish communist governments and to join Soviet-dominated organizations like the economic mutual-aid organization COMECON (established in 1949 in opposition to the American Marshall Plan to aid post-war recovery in Europe), and the Warsaw Military Pact (1955, in opposition to NATO) (see figure 2.3). Soviet-style political systems and the Soviet development model were imposed upon them. Attempts by some of these countries to reassert their independence, as with Hungary in 1956 and Czechoslovakia in 1968, were brutally repressed (Soviet determination to maintain the status quo in Eastern Europe was dubbed in the West the Brezhnev doctrine). In time, the Soviets realized the necessity of allowing some limited autonomy to these countries, most notably Hungary with its special model of the command economy, and Romania under the idiosyncratic and repressive Ceauşescu dictatorship. However, they were cut off from the rest of Europe by Soviet suspicions and by the Cold War. The accompanying arms race added enormously to international tensions in the years after 1945.

Outside Eastern Europe, Soviet influence and the spread of communism did not always go hand in hand. The communist victory in China in 1949 was a great boost to Soviet morale, and for a number of years the Soviets exercised considerable influence in that country. By the 1960s, however, relations had cooled considerably as a result of rival interpretations of the communist creed, competition over the leadership of the communist world and territorial disputes. Towards the end of that decade the rivalries led to the threat of open warfare. Communist takeovers in some other parts of Asia, and liberation and revolutionary movements elsewhere, often attracted Soviet support, though usually in an indirect form. Meanwhile, the Soviets did their best to make friends in various parts of the Middle East, south Asia and other regions, while the growth of Soviet naval power and commercial activity from the

1960s was likewise designed to boost their influence and to weaken the capitalist West. In the latter aim, however, these policies were largely unsuccessful.

In official Soviet propaganda, the spread of communism and of Soviet influence were the inevitable consequence of the victory of socialism over a disintegrating capitalism. In many parts of the non-communist world, however, and especially in the United States, they were regarded as manifestations of Soviet imperialism. Eastern Europe especially tended to be regarded as part of the Soviet empire (a view shared by many East Europeans) – albeit, perhaps, an empire of political domination rather than of economic exploitation in the traditional view of imperialism. The fear was that this imperialism might spread into neighbouring regions and ultimately lead to world domination by the USSR. From the late 1940s, therefore, the USA and its allies pursued a policy of 'containment', under the inspiration of the American diplomat George F. Kennan. 'Containment' meant the attempt to prevent Soviet expansionism by surrounding it on all sides by military bases and allies friendly to the Western cause. Kennan was in fact influenced in his thinking by the geopolitical ideas of the British geographer Halford Mackinder. Earlier in the twentieth century Mackinder had taught that any state dominating what he called the 'Heartland' or 'geographical pivot' (the central part of the Eurasian continent, with its defensive capabilities and wealth of resources) was well placed to dominate the world. Since the USSR already ruled the 'Heartland', Kennan thought it must be prevented from breaking out from its 'fortress' base by a policy of siege. Mackinder's and Kennan's ideas were subsequently much criticized. As Taylor has noted, Mackinder's ideas seemed dated by the second half of the twentieth century, and in any case geopolitical thinking has generally served an ideological purpose – to justify particular political policies rather than to provide objective analysis (Taylor, 1993, pp. 53–64). The 'Heartland' theory found little echo in Soviet thought. The eminent Soviet geographer Yu. G. Saushkin regarded Mackinder as having given a particularly reactionary interpretation to the geopolitical ideas of the German geographer Friedrich Ratzel, and as indulging in 'speculation about the geographical dependency of the world strategy and policy of various states' (Saushkin, 1973, p. 75).

While the Soviets may not have thought much of 'Heartland' theory, they certainly did all in their power to turn their empire into a fortress by spending huge amounts of money on its defence and attempting to seal it off as far as possible from the outside world. Only in the late 1960s was this policy eased somewhat by the need to enter the world market to earn much needed hard currency. Some scholars have interpreted this event as the USSR being absorbed by the single world economy as a semi-peripheral state (see Bradshaw and Lynn, 1996). Be that as it may, problems in the Soviet economy and growing contacts with the world outside made the fortress mentality increasingly difficult to sustain. By 1980 the Soviets were having problems in maintaining control in Poland. They also invaded Afghanistan in an attempt to sustain a communist government there, even though that country lay outside the post-war Soviet bloc. Gorbachev's domestic policies after 1985 were de-scribed in chapters 3 and 4. His foreign policy was aimed at easing tensions with the West, thus permitting disarmament and economic reconstruction. It may well be, as some scholars have argued, that Gorbachev's policies were ultimately designed to reassert the USSR's status as a superpower. But his domestic failures plus his refusal to risk further confrontation with the West undermined the Soviet position in Eastern Europe. By the time the USSR itself broke up it had already lost its former empire in Eastern Europe and withdrawn from Afghanistan. The stage was set for the adoption of a new Russian policy towards the outside world.

Yet that policy must be influenced to a degree by the Soviet past. Some of the most obvious ways in which this has happened to date have included the need to sort out problems left by the Soviet regime, such as the question of Soviet international debt, Soviet property and the difficulties resulting from the stationing of former Soviet troops in other countries. Other issues are longer term. One is the inevitable tensions and suspicions which must remain after the many years of confrontation with the West, China and other areas. The fact that Russia is the USSR's most direct successor state, heir to most of its territory, resources and military power, leads to the suspicion that it may also be heir to some of its ambitions. Fear of a reassertion of Russian imperial-ism is by no means confined to eastern and central Europe. Russian fears of Western and Chinese expansionism are equally real on the other side. Memories are not soon relinquished, and the fact that both in Russia and abroad there are many personalities and vested interests with Cold War attitudes may ensure their survival for quite some time. Another issue is the territorial claims and counter claims left over from the Soviet era. These must influence Russia's relations with some of its neighbours. Finally, the legacy of the arms race, with a huge military capability and a military–industrial complex still employing millions of people and producing enormous quantities of arms, will inevitably constrain Russian policy options towards the outside world and mould the way the world views Russia.

Post-1991 Influences on Russian Policy

Whatever may be the Soviet heritage with which the new Russia must cope in cultivating its relations with the outside world, the collapse of the USSR inevitably means new challenges and new priorities. No longer can Russia claim the status of superpower, and this fact alone has far reaching implications for foreign relations. Moreover, a marketizing and democratizing Russia will necessarily have different goals from an authoritarian Soviet Union. Even so, it must be acknowledged from the outset that different groups and individuals in Russia will have different ideas of what those foreign policy goals should be. The ideas of some are not terribly different from those which reigned in the former Soviet Union. One of the problems of foreign policy development in the new Russia has been the fact that different institutions and groups have to some extent been following their own, often contradictory, policies. This is a reflection of the diminution of central authority.

One factor which must certainly help to govern Russia's relationships with the outside world is the economic one. Marketization implies an openness to the world economy for the sake of efficiency. Arguably, one reason why the old Soviet Union lagged behind was the barriers which were erected to trade and to the free flow of ideas and people. A Russia eager to modernize its economy must be prepared to be open to global economic forces to attract international capital and to sell its own goods and services abroad. An economy closed to the outside world runs the risk of developing or preserving economic activities which are uncompetitive in the world market place. This was the problem which Soviet industry faced when it tried to export. And it is the problem which a great deal of Russian industry faces today. Of course, the degree of openness in an economy is a matter of political judgement about which opinions will vary. Some in Russia, like those associated with the 'shock therapy' policies of the early Yeltsin years, favour openness on the grounds that it is better to be competitive than to preserve industry and jobs. Others, like the communists or the Civic Union of industrial interests, take a different view. The world trend, however, is presently towards openness, and Russia is under much pressure from international agencies like the International Monetary Fund (IMF) and the World Bank, as well as from other states to move resolutely in this direction. These days, of course, Russia is a conglomerate of federal, regional and local interests, and this feature will also have a bearing on the evolving situation.

The age old security problem will continue to be a major factor moulding Russia's developing relationship with the outside world. As earlier chapters of this book have shown, the military factor has loomed large in Russian history. It is impossible to imagine its playing a minor role in the near future. Anxieties about a future potential threat from the West (mirrored in some Western attitudes towards Russia, which, as the opening quote to this chapter suggests, have a remarkably long pedigree), uncertainty about possible threats to security in central and eastern Europe, the presence of volatile states along Russia's long southern border and its proximity to China in the east all ensure that there will be a continuing concern with security – the more so since there are many vested interests from the Soviet 'imperial' past still eager to preach the need for it (Kontorovich, 1996). Russia's determination to preserve its territorial integrity and internal order will also affect foreign affairs. Whether those who advocate a more aggressive and even imperialistic stance in foreign affairs will be able to increase their influence remains to be seen.

In the last analysis, the way Russians view the outside world reflects the way they view themselves. This is perhaps the most controversial issue of all, since it concerns not only how Russia should relate to other countries but also what sort of country Russia itself should try to become. In Soviet days, such questions were determined by the official ideology, as interpreted by the political leadership. These days there is much more scope for a plurality of political opinions, as represented in public debate between different groups and political parties. Tuminez, for example, argues that public debate since Gorbachev's day has been structured around variants of nationalism (Tuminez, 1996). She categorizes nationalists into four types: *liberal nativists*, who define Russia's mission as 'primarily defensive and inward-looking: to restore and defend the physical and spiritual well-being of the land and people'; *Westernizing democrats*, who regard Russia as 'a great power whose immense potential can be realised through democracy, market reform, and integration and participation in the international community'; *statist nationalists*, who tend to reject communism while deploring the disintegration of the Soviet Union, and advocate a larger, leading role for Russia in the former Soviet Union and a foreign policy that 'distinctly defends Russian national interests'; and, finally, *national patriots* of a variety of leftist and rightist views, including Stalinists and fascists, who generally want to see Russia restored to its former status as a great imperial power. The latter two positions, and particularly the last, thus advocate foreign policies which would inevitably increase international tensions.

From a geopolitical perspective, the debates are sometimes conceptualized as being between *Atlanticists* and *Eurasians* (Rahr, 1992). Put this way, the debates to some extent reflect those between Westernizers and

Slavophils in the nineteenth century (see chapter 1). Thus Atlanticists (like President Yeltsin and his associates, especially before 1993) tend to view Russia as part of the European and Western family of states – in other words, as very much a part of the affluent northern countries of the world rather than of those in the developing south. Atlanticists tend to advocate marketization, democratization, liberal economic policies and the adoption of Western cultural values. By contrast, the Eurasians tend to argue against too much Westernization and warn that Russia neglects its longstanding and unique relations with its Islamic and southern neighbours at its peril. Sometimes Eurasians argue that Russia has a unique, integral civilization, different from that of Europe, and emphasize that it occupies a geopolitical position astride Europe and Asia. It can thus act as a bridgehead between the two continents. While the two positions are more a matter of emphasis than of substance, the Atlanticists generally advocate policies which would find greater support in the West.

The remainder of this chapter will consider specific facets of Russian foreign relations as they have developed since 1991.

International Economic and Trade Relations

Although Soviet foreign trade began to grow quite quickly in the 1970s, it remained insignificant by world standards. In 1985, for example, the Soviet Union generated 15 per cent of global GNP but contributed only 3 per cent to world trade (Bradshaw, 1995, p 132). One reason for this was that until 1987 foreign trade remained a state monopoly. At the beginning of that year, the Soviet Ministry of Foreign Trade lost the monopoly over trade with other countries which had first been established in April 1918. Ministries, associations and other organizations were now granted the right to engage in foreign trade, and by 1989 that right had been extended to the level of the enterprise. Even so, central government continued to claim a large share of the revenues generated by foreign trade, and many restrictions and controls remained in force. As Bradshaw argues, control over the earnings of foreign trade, and particularly over those generated by the export of energy and other natural resources, became central to the tussle between the centre and the regions and republics which marked the end of the Gorbachev years. Other innovations of this period, like the establishment of joint ventures and the creation of special economic zones, also had only a minimal effect in opening the economy up to the outside world. In any case, foreign trade suffered greatly as a result of the general collapse of the Soviet economy.

The opening up of the Russian economy to world trade was an important part of the Yeltsin economic reforms which took effect in 1992 (see chapter 4). Finally abolishing the state monopoly on foreign trade was seen as essential to marketization, the efficient restructuring of the economy and the attraction of inward investment. Henceforward, controlling imports and exports was to be done mainly by means of tariffs and subsidies rather than by direct state controls as in the past. Pressure from the IMF and other international agencies was in the direction of doing away with measures of protection and similar policies which discriminated against foreign businesses. The indirect subsidies which CMEA members had enjoyed, especially in energy supplies, had come to an end in 1990, when the CMEA itself effectively collapsed. From 1991, trade between former CMEA members was conducted in hard currency. Russia's trade with the former CMEA thus fell even more quickly than it did with the rest of the world.

Within the former USSR, Russia continued to subsidize the other republics by selling them energy at below world market prices in 1992 and 1993. There was also a variety of credit arrangements. Later, energy sales approached world market prices but Russia in effect continued to subsidize its neighbours by supplying energy despite their growing indebtedness (Kontorovich, 1996, pp. 24–5). Even so, trade between Russia and the rest of the CIS fell precipitately as a consequence of the collapse of the command economy. Despite attempts to establish a common economic space through the CIS, the ruble zone fell apart. Increasingly, it became clear that Russia's future trade links would be less with its neighbours in the 'Near Abroad', or with the former Soviet satellites in Eastern Europe, than with the developed world. The policies of the Yeltsin government encouraged this trend. Yet at the time of writing, Western investment in Russia (by way of joint ventures but now increasingly by other means such as direct investment) has been relatively modest. Corruption, Russian taxation policies, red tape and legal uncertainties have been blamed for this reticence. The problems in the Russian economy and some aspects of trade policy are also blamed by commentators for the refusal to date to admit Russia into membership of the World Trade Organization and the Group of Seven.

The period since 1991 has therefore seen a remarkable reorientation of the geography of Russian foreign trade. In 1990, for example, 70 per cent of Russia's exports went to other Soviet republics, which in turn provided 47 per cent of its imports. In 1996, the CIS took 21 per cent of Russian exports and provided 30 per cent of its imports. Such figures should be treated with some caution because of the price changes which have accompanied the transition to market relationships, but they do give some idea of the switch in emphasis. Russian trade with the former

CMEA countries, which was running at about 43 per cent of exports and 42.5 per cent of imports of Russia's non-CIS trade in 1990, had fallen to 20 and 17.3 per cent respectively by 1992 (Bradshaw, 1995, p. 140). In 1995, trade with the six European former CMEA members was running at about 13.8 per cent of non-CIS exports and 10.6 per cent of equivalent imports. Trade with the EU, however, reached almost 40 per cent of non-CIS exports and 54.3 per cent of non-CIS imports by 1995. Altogether, the developed OECD nations accounted for 61 per cent of Russia's non-CIS exports in 1995 (compared with 37.4 per cent in 1990) and 70 per cent of its equivalent imports (38.3 per cent in 1990). Russia's strongest trading partners in 1995 were: for exports, Ukraine (8.5 per cent of total), Germany (7.6 per cent), the USA (5.4 per cent), Switzerland (4.4 per cent), China (4.2 per cent), Italy (4.1 per cent), the Netherlands (4.0 per cent), the UK (3.9 per cent), Japan (3.9 per cent) and Belarus' (3.7 per cent); and for imports, Ukraine (14.2 per cent), Germany (14 per cent), Kazakhstan (5.9 per cent), the USA (5.7 per cent), Belarus' (4 per cent), Italy (4 per cent), the Netherlands (3.5 per cent), Poland (2.8 per cent) and the UK (2.4 per cent). These data do not include shuttle trade.

Russia's most important exports in 1996 were mineral products, including oil and gas, accounting for 46.9 per cent of exports to non-CIS countries, and metals, accounting for 26.4 per cent. Other significant exports included chemicals (8.2 per cent) and machinery and equipment (7.7 per cent). The most important imports from equivalent countries included machinery and equipment (37 per cent), foodstuffs (24.9 per cent) and chemicals (15.8 per cent). Russia is thus locked into a trading pattern which is characteristic of developing countries, exporting raw materials and importing finished products. This reflects the uncompetitive nature of its industry. It also suggests why, in the new trading environment, it is the resource-rich regions of Russia which have the greatest political advantages.

The emerging patterns of foreign trade, then, suggest strong economic reasons why it may be in Russia's interests to maintain good political relations with the West generally, while also seeking to maintain its links with its neighbours in the 'Near Abroad'. As we know, however, economics do not always determine politics. The remainder of this chapter will highlight some of the other factors which are at work.

Russia, Europe and the West

With the collapse of the Soviet Union in 1991, the European territory now under Moscow's control has retreated markedly to the east. Not

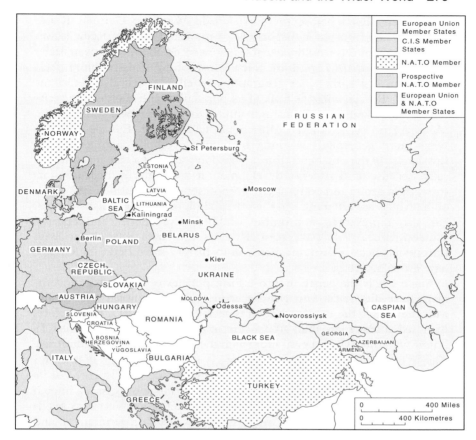

Figure 11.1   Russia and the states to the west.

only are the former eastern European countries now liberated, but the western republics of the former USSR have also gained their independence. Russia's links with the rest of Europe now lie across a broad belt of separate and partially antagonistic states, or via a much reduced sea coast (figure 11.1).

Many observers in the former Soviet republics and the erstwhile Soviet bloc states, and some in the West, have expressed the hope that this signals a real disengagement by Russia from Europe. The hope is that Russia, whose power has so often overshadowed the rest of the continent and especially its eastern part, will now become so preoccupied with its own internal affairs, or perhaps with opportunities farther east, that it will leave the rest of Europe alone. In fact this idea is by no means new. Even before the First World War, the German Kaiser Wilhelm II encouraged Tsar Nicholas II to make Russia a Pacific power in the hope that this would make it less concerned about European issues (Seton-Watson, 1967, p. 586). Such a turn of events is no more likely today than it was then, for geography as well as much else determine otherwise. Russia, with the bulk of its population and economic activity west of the Urals, simply cannot afford to turn its back on Europe. Moreover, as earlier parts of this book have argued, Europe (and the West more generally) has long had deep symbolic significance for Russians. On the one hand, Europe has long seemed a threat: economically, because of its superior development level; militarily, because of its strength in arms; and culturally, – as representing an outlook and a way of life which are different. On the other hand, to many Russians Europe has also represented a promise: of liberty, prosperity and cultural achievement. As we have seen, Russia's love–hate relationship with Europe has been a constant theme in its history since Peter the Great's day, if not before.

Since 1991, the generally pro-reform policies of the Yeltsin government have dictated a much closer economic and political relationship with the rest of the continent, and especially with the EU, than before. As noted above, more than 40 per cent of Russia's trade outside the CIS is now with the EU, and even more with the developed OECD countries. The West, and the EU in particular, is seen as a major market for Russian products and a significant source of future investment. For its part, the West understands the potential importance of Russia's huge if underdeveloped and somewhat unstable market. The West also needs Russian cooperation over a range of issues, from disarmament and nuclear power to the environment, organized crime, drugs trafficking and money laundering.

There is much, then, to draw Russia and the West closer together. Equally, however, there are numerous problems and suspicions which remain from the past. For its part, Russia continues to be wary of the West's military potential and fears the hawkish sentiments of some circles in the former Eastern Europe, the EU and the United States. There is a traditional Russian resentment of being denigrated in the West, especially now that Russia can no longer be regarded as a superpower, or of being shut out of European developments. Some Russians also express concern about the insidious effects on their society of Western culture, of creeping Americanization. For its part, the West fears a Russian return to its bad old ways, and especially the future advent of a more totalitarian government with a propensity for militarism and an imperialistic foreign policy. The problem is that, with its current economic and political difficulties, Russia can clearly not be regarded as a normal member of the Western community and it may not be possible to treat it so in the near future without endangering the West's own stability. Equally, since 1991, it is no longer possible to regard Russia as 'the Other'. Part of the difficulty of deciding exactly how to treat Russia lies in the uncertainties which surround the question of how Russians regard themselves.

The existence of the former eastern European states (now frequently referred to as central European) only complicates the issue. As we have seen, after 1945 most of these formed the USSR's *cordon sanitaire* against the West but, having escaped communist domination in 1989, they are now understandably determined not to fall into Moscow's clutches again. There is an anti-Russian legacy in many of these states which is only reinforced by the activities of émigrés in western Europe, the United States and elsewhere. Yet, in a longer historical perspective, the attitude towards Russia has been by no means uniform throughout this region – there are, after all, numerous links in the Slavic languages and culture, in the Orthodox faith and in other factors to bind Russians with at least some of their western neighbours, despite the many misfortunes of the past. Historic Russian links with Serbia and some of the other Balkan countries surfaced once again in the 1990s, especially in the Yugoslav crisis, in ways not always pleasing to the Western allies.

Many Russians, and not only those of a nationalistic or communist persuasion, have expressed fears that, having abandoned its strategic position in central and eastern Europe after 1989, Moscow now faces the possibility that these same regions will be used by the West to threaten its security. Events in the mid-1990s seemed to confirm these fears. Many of the ex-communist states (including some of the former Soviet ones) applied to join NATO, partly out of an understandable fear that a resurgent (and nuclear-armed) Russia might seek to overawe them once again. A number of Russia's policies, to say nothing of Duma resolutions and declarations, seemed to confirm their fears, including Russia's determination to protect its ethnic minorities in the 'Near Abroad', its activities in Chechnya, Abkhazia and Crimea (including Sevastopol') and the possibility that it might use its forward positions in Kaliningrad Oblast and the breakaway Transdniester Republic in Moldova to increase its military influence in the region. Negotiations for a union with Belarus' also seemed part of this pattern (for all these issues, see chapter 10). Tensions between Russia and the West increased when the USA and its allies expressed their support for NATO expansion, but these may now be eased following a security agreement between Russia and NATO in 1997. Prospects for EU expansion, including the likely accession of Poland, the Czech Republic and Hungary to membership in the near future, led to fears that this too was helping to isolate Russia. The latter's

'Partnership for Peace' agreement with NATO in 1994 and its Partnership and Cooperation Agreement with the EU may go some way to allay such fears.

After 1917, and particularly after 1945, the USSR's great ideological as well as military foe was the United States. The collapse of the USSR has done much to heal old wounds, but residual suspicions remain, especially since the geopolitical positions of the USA and Russia mean that they must meet in various arenas in addition to the European one, and that their interests are unlikely to be identical. The remainder of this chapter will consider two other arenas especially important to Russia: the Middle East and the Far East–Pacific Basin area.

Russia and the South

In the Soviet period, the USSR regarded the regions to its south as both a potential opportunity and a potential danger. From the days of Lenin, Soviet leaders hoped and worked for the spread of communism in the region, in the belief that this would demonstrate to colonized or exploited peoples throughout the world how to throw off the yoke of capitalism. Later in the Soviet period, they tried to use the political instabilities in the region to defeat Western policies of 'containment'. At the same time, the Soviets worried that the West might use the region's political volatility to destabilize the USSR and that the spread of militant Islam might have the same effect.

Since 1991, with the independence of the Transcaucasus and Central Asia, Moscow is much less immediately concerned with events in the area than before (figure 11.2). However, as noted in Chapter 10, the existence of politically unstable states along its southern boundary, plus the possibility that the region might fall under the influence of rival powers, is of concern, and the Russians have been trying to influence events and secure the former Soviet frontier against incursions by destabilizing elements. One very important consideration influencing Russian attitudes is how events in the Islamic world might affect Russia's own very sizable Muslim population. The spread of militant Islam is seen as a possible security threat, while there is concern lest the adoption by Moscow of an anti-Muslim stance in its foreign affairs might encourage unrest among its own Muslim subjects. Clearly, such considerations had only minimal influence in the case of the Chechen War, when Moscow launched an attack against Muslim separatists. In this case it was evidently felt that the dangers of a successful secession outweighed those of Muslim unrest elsewhere in Russia (to what extent civilian policy-makers in Moscow were in charge of events in Chechnya is

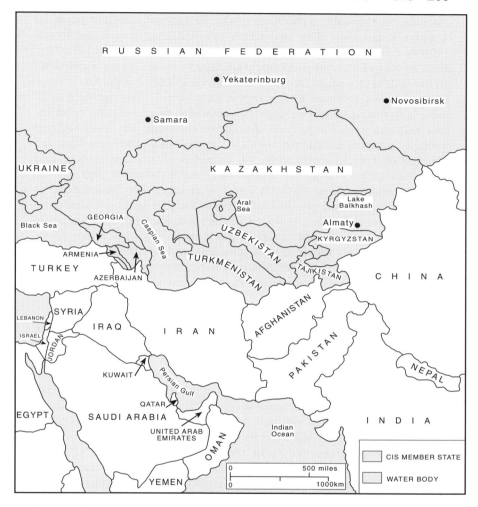

Figure 11.2 Russia and the states to the south.

uncertain). Reluctance by Moscow to endorse Western policies in
the Middle East, the Balkans and elsewhere which might be regarded
as anti-Muslim clearly do relate to Russia's peculiar geopolitical
circumstances. These circumstances are cited by some of the so-called
'Eurasianists' as reasons for urging caution on Russian leaders considering following the West's foreign policy lead.

In the Soviet period, Moscow deliberately cultivated its relations with
the Arab world and with India as a way of embarrassing the West; for

example, in the latter's handling of the Palestinian question or in its dealings with China. The Soviet invasion of Afghanistan at the end of 1979 to bolster its communist government was viewed by some in the West as part of a grand geopolitical strategy to secure warm water ports on the Indian Ocean and menace Western oil interests in the Gulf. Since 1991, Russia's activities in this region have been less obviously part of some larger expansionist strategy. Nevertheless, there are some continuities with the past: the pursuit of security, the quest for resources and markets and more nebulously the desire to maintain influence. Thus the securing of Russian access to the oil resources of the Caspian Sea, and the policy to encourage the export of Central Asian oil and gas via Russian territory rather than via some other route, can be regarded as furthering Russia's geopolitical influence in the region. The same can be said of the policy to contain the expanding influence of other powers in Central Asia, notably that of China and Turkey. Recent Russian drives to cultivate warmer relations with Iraq, Iran and China (and the selling of arms to India) are related to Russia's security and also its economic interests, though such policies are by no means always pleasing to the West.

Russia, the Far East and the Pacific Basin

Like the United States and Canada, Russia's geographical location enables it to play a role in both Atlantic and Pacific regions. In Russia's case, however, the distance separating its Pacific regions from its developed core in the west is particularly great. Some of the problems of the Russian Far East, deriving from its remoteness and underdevelopment, were discussed in chapter 8. In the Soviet period, official priorities included developing the region as a resource supplier for the rest of the USSR (and to much lesser extent for export) and exploiting its strategic potential. The region's military role loomed large not only because of the long land frontier with China but also because it enabled the Soviets to have a presence in the North Pacific, thus challenging US predominance in this area. Furthermore, the Far Eastern ports were the means whereby the USSR maintained its military and economic links with various client states like Vietnam and North Korea. Unfortunately, as chapter 8 suggested, this Soviet legacy has left Russia's eastern periphery ill equipped for the new role it must play in the post-communist era, with an infrastructure badly in need of renewal and an exaggerated military dependency which could no longer be afforded even if it were needed. The changes of the post-Soviet period have limited Moscow's ability and probably its willingness to provide economic help to the region, and have

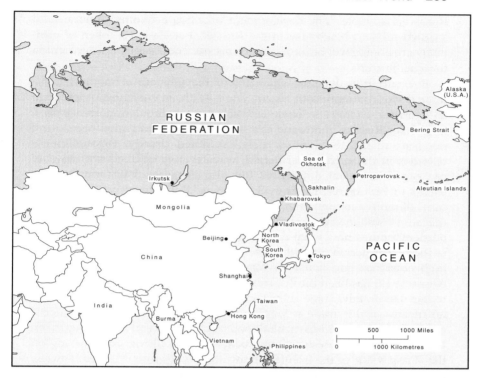

Figure 11.3   Russia, eastern Asia and the Pacific.

thrown it much more on its own devices. Little wonder that this has led to an upsurge of autonomous sentiment in this most remote of territories.

That said, Russia's Far Eastern territories are of enormous potential significance to the country, not only because of their resource endowment but also because of their location close to some of the world's most dynamic economies (figure 11.3). In 1995, according to the not always reliable official statistics, about 22 per cent of Russia's non-CIS exports and 17 per cent of its imports were with countries bordering the Pacific Ocean (including the USA). About 13 per cent of Russia's non-CIS trade turnover was with the so-called Asian-Pacific states, the industrial economies situated on the western side of the Pacific Ocean. The latter statistic represents hardly any proportional increase from the late Soviet period, mainly because of the Far Eastern region's inherent problems and its remoteness from European Russian industry. In the future, however, the situation may change to the advantage of this region and

Russia as a whole. The development of a free economic zone around Vladivostok and Nakhodka in the Russian Far East is a token of what many are hoping will become a much greater commercial involvement in the Pacific arena.

The currently dynamic character of the Chinese economy has attracted world attention in recent years. With its enormous population and territory, China is obviously of great potential importance as a market for Russian goods and as a supplier of food and other necessities, not least to the Russian Far East. As noted already, Russo-Chinese relations in the communist period were severely strained not only because of ideological differences but also because of historic Chinese claims to various territories in Soviet Asia. Both sides spent enormous sums of money strengthening their frontier and strategic defences. From the late 1980s, however, relations have begun to improve and most of the border disputes now seem to have been resolved. A remarkable feature of the recent period, one which reflects economic changes under way in both countries and which in many ways harks back to the situation before 1917, has been the flourishing of cross-border trade and contacts, to the obvious advantage of the southern part of the Russian Far East. By no means all this trade is legal, however, and there is much concern on the Russian side in particular about the incidence of illegal migration. Up to 200,000 Chinese are now believed to be living and working on the Russian side of the frontier, many illegally. Future relations between the two states will obviously depend among other things on the political evolution of China. Some Western observers fear that a future, more nationalistic Russia might be tempted to move closer to China as a way of breaking out of its international isolation. This might lead to a resurrection by Western states of the 'containment' policy outlined above. Equally, however, a future more powerful and more aggressive China might persuade the West to try to use Russia as a counterweight. In 1995, according to official data, China took about 5.2 per cent of Russia's non-CIS exports and provided around 2.6 per cent of its imports, though much trade goes unreported.

Moscow's relations with Japan after 1945 were made difficult not only because of the latter's dependency on the USA but also because of Japanese claims to several islands in the southern part of the Kuril chain which were occupied by Soviet troops at the end of the Second World War and never returned. In the 1970s, Japan became involved in a number of compensation agreements with the USSR, receiving coal, timber and fish in exchange for help in developing Soviet eastern resources. Further development of this trade, which might have benefited both sides, was held back by the territorial dispute. This so-called 'Northern Territories Dispute' has continued to complicate relations between Russia and Japan since 1991, the Russians evidently placing great weight on the security advantages deriving from continued occupation of the disputed islands. The issue is exacerbated by disagreements over fishing rights in the Sea of Okhotsk and neighbouring waters. While Japan is certainly interested in the economic opportunities presented by the Russian Far East, territorial and other disagreements (for example, over the Russian dumping of nuclear waste) seem likely to place limits on the further development of relations in the near future. In 1995, Japan accounted for about 5.5 per cent of Russia's non-CIS exports and 2.3 per cent of its imports.

While Russia's relations with some of the USSR's former client states like North Korea have cooled since 1991, those with old-time capitalist foes like South Korea have warmed considerably. Assuming that the present economic difficulties facing the Asia Pacific economies are only temporary, it seems hard to believe that Russia's economic linkages with most of them will not continue to grow, despite current economic difficulties in the region.

In the Soviet era, the entire northern Pacific and the Arctic Ocean to the north were potential zones of conflict in a future war between the USSR and the USA, and this fact had a profound influence on the social and economic character of adjacent territories. With the end of the Cold War, Russia seems set to transform itself into a Pacific power of a much more positive kind and to participate in the expanding Pacific basin economy. As yet, however, this is very much a promise for the future. The depressing legacy of the Soviet past seems likely to have a continuing influence on the region for some time to come.

Conclusion

Since the collapse of communism, it is certain that Russia is no longer a superpower, yet it can hardly avoid being a great one. The largest country in the world, with a huge population and vast resources, it spans the Northern Hemisphere. Russia has a geographical presence in two of the world's great oceans, overlooks the Arctic and looms just to the north of the Middle East. This unique geopolitical position virtually guarantees Russia influence over world affairs. But it also guarantees that the world will be interested in Russian ones. As it adjusts to its new position in the world and moves away from decades of isolation, Russia is faced with many opportunities and inherits a multitude of problems. Whether it remains a prisoner of the past or can face up to the challenges posed by geography, history and a world in precipitate change will determine if it can at last find the security and prosperity which have eluded it for so long.

Some Place Name Changes in the Post-Soviet States

Post-1991 name	Soviet name
Post-Soviet states	
Belarus'	Belorussia
Kyrgyzstan	Kirgizia
Moldova	Moldavia
Russian Federation/Russia	RSFSR
Autonomous territories of the Russian Federation	
Bashkortostan	Bashkiria
Kalmykia-Khal'mg-Tangch	Kalmykia
Sakha (Yakutia)	Yakutia
Tatarstan	Tatar ASSR
Tyva	Tuva
Cities in the Russian Federation	
Nizhniy Novgorod★	Gor'kiy
St Petersburg★	Leningrad
Samara★	Kuibyshev
Tver'★	Kalinin
Vladikavkaz★	Ordzhonikidze
Yekaterinburg★	Sverdlovsk
Cities outside the Russian Federation	
Almaty (Kazakhstan)	Alma Ata
Ashgabat (Turkmenistan)	Ashkhabad
Bishkek (Kyrgyzstan)	Frunze
Chisinau (Moldova)	Kishinev
Kyiv (Ukraine)	Kiev

★ Pre-1917 names

Glossary

ASSR	Autonomous Soviet Socialist Republic
autonomous oblast (AOb)	autonomous region
autonomous okrug (AO)	autonomous district
Bolsheviks	early name for the Russian communists (or more properly that fraction led by V. I. Lenin)
cossacks	originally, refugees and others leading a semi-nomadic lifestyle on the steppe; they later took on a military role for the Russian state, particularly as frontier guards.
CIS	Commonwealth of Independent States
CMEA	Council for Mutual Economic Assistance (also COMECON); the economic organization which facilitated trade between the Soviet Union and its communist allies
Duma	the lower house of the Russian parliament
Gazprom	Russia's privatized monopoly gas producer
GDP	gross domestic product
GNP	gross national product
Goskomekologiya	the state committee for environmental protection
Goskomstat	the official statistical service of the Russian Federation
Gosplan	the (Soviet) state economic planning committee

ILO	International Labour Organisation
IMF	International Monetary Fund
KGB	Komitet Gosudarstvennoy Bezopasnosti (State Security Committee), the Soviet secret police
kolkhoz	collective farm
kray	territory
NATO	North Atlantic Treaty Organization
NEP	New Economic Policy (*c.* 1921–8)
NKVD	Narodniy Komissariat Vnutrennikh Del (People's Commissariat for Internal Affairs), the ancestor of the KGB
nomenklatura	term referring to privileged elements and important officials in the Soviet period
oblast	region
okrug	district
perestroyka	literally 'reconstruction', the slogan used to describe the reforms of the Gorbachev period (1985–91)
rayon	district
RSFSR	Russian Soviet Federal Socialist Republic, the official name of the Russian Federation in the Soviet period
soviet	a legislative council or parliament, operating at national, republican, regional or local level; after 1917, soviets provided the official framework for communist government, hence the origin of the name USSR, or Soviet Union
sovkhoz	state farm
Union republic	one of the principal republics, like Russia or Ukraine, that were combined to form the USSR
USSR	Union of Soviet Socialist Republics (or Soviet Union)

Further Reading

Because Russia continues to be a region of rapid change, students are advised to consult newspapers, current affairs journals and other media as much as possible. There is a growing number of newssheets of various kinds available. The following will be found to be particularly useful:

Current Digest of the Post-Soviet Press. Weekly with quarterly index. Translated press and current affairs articles.

Post-Soviet Geography and Economics (previous titles: *Soviet Geography*, *Post-Soviet Geography*). Currently ten issues per year. Geographical and economics articles, mainly by Western scholars, with an accent on the present day.

Transition. Open Media Research Institute, Prague. Fortnightly. Current affairs articles, mainly by Western journalists and specialists.

Economist Intelligence Unit. Country Reports. Regular reports on political and economic developments in Russia and the other post-Soviet states. Reports are issued for each of the states either individually or in groups which change from time to time. There are also reports for the CIS.

Post-Soviet Affairs. Quarterly. Mainly devoted to current economic and political issues.

Websites are growing in importance for students of Russia and the other post-Soviet states. Some useful sites for Russia are:

ABSEES Online. http://carousel.lis.uiuc.edu/-absees/search.html
OMRI – Open Media Research Institute. http://www.omri.cz/Index.html
REESweb: Russian and East European Studies, University of Pittsburgh. http://www.pitt.edu/-cjp/rees.html
RFE/RL Newsline. http://www.rferl.org/newsline
Russia on the Net. http://www.ru/
St Petersburg Times. http://www.spb.su/times/
TACIS. http://www.nns.ru/analytdoc/otch1.html

General geography

Bater, J. H. (1996) *Russia and the Post-Soviet Scene: a Geographical Perspective.* London: Arnold.
Bradshaw, M. J. (ed.) (1991) *The Soviet Union: a New Regional Geography?* London: Belhaven.
Bradshaw, M. J. (ed.) (1997) *Geography and Transition in the Post-Soviet Republics.* Chichester: Wiley.
Lydolph, P. E. (1990) *Geography of the USSR.* Elkhart Lake, WI: Misty Valley.
Shaw, D. J. B. (ed.) (1995) *The Post-Soviet Republics: a Systematic Geography.* Harlow: Longman.

Physical geography and resources

Barr, B. and Braden, K. (1988) *The Disappearing Russian Forest.* Totowa, NJ: Rowman and Littlefield.
Knystautas, A. (1987) *The Natural History of the USSR.* New York: McGraw-Hill.
Lydolph, P. E. (1977) *World Survey of Climatology, Volume 7. Climates of the Soviet Union.* Amsterdam: Elsevier.

Environmental disruption and conservation

Feshbach, M. and Friendly, A. (1992) *Ecocide in the USSR: Health and Nature under Siege.* London: Aurum.
Massey-Stewart, J. (ed.) (1992) *The Soviet Environment: Problems, Policies and Politics.* Cambridge: Cambridge University Press.
Panel on the State of the Soviet Environment at the Start of the Nineties (1990) *Soviet Geography*, 31(6), 401–68.
Peterson, D. J. (1993) *Troubled Lands: the Legacy of Soviet Environmental Destruction.* Boulder, CO: Westview.
Pryde, P. R. (1991) *Environmental Management in the Soviet Union.* Cambridge: Cambridge University Press.
Pryde, P. R. (ed.) (1995) *Environmental Resources and Constraints in the Former Soviet Republics.* Boulder, CO: Westview Press.

History and historical geography

Bater, J. H. and French, R. A. (eds) (1983) *Studies in Russian Historical Geography.* 2 volumes. London: Academic Press.
Davies, R. W., Harrison, M. and Wheatcroft, S. (1994) *The Economic Transformation of the Soviet Union, 1913–45.* Cambridge: Cambridge University Press.

Dukes, P. (1998) *A History of Russia: Medieval, Modern, Contemporary c.882–1996*, 3rd edn. London: Macmillan.

Hosking, G. (1992) *A History of the Soviet Union, 1917–1991*, final edition. London: Fontana.

Nove, A. (1992) *An Economic History of the USSR*, 3rd edn. London: Penguin.

Pallot, J. and Shaw, D. J. B. (1990) *Landscape and Settlement in Romanov Russia, 1613–1917*. Oxford: Clarendon Press.

Economy and economic geography

Ellman, M. and Kontorovich, V. (eds) (1992) *The Disintegration of the Soviet Economic System*. London: Routledge.

Gregory, P. R. and Stuart, R. C. (1998) *Russian and Soviet Economic Performance and Structure* 6th edn. New York: HarperCollins.

Shabad, T. (1969) *Basic Industrial Resources of the USSR*. New York: Columbia University Press.

Agriculture and rural geography

Channon, J. (1995) *Agrarian Reforms in Russia, 1992–95*. London: RIIA.

Ioffe, G. and Nefedova, T. (1997) *Continuity and Change in Rural Russia: a Geographical Perspective*. Boulder, CO: Westview.

Planning and urban geography

Andrusz, G. D. (1984) *Housing and Urban Development in the USSR*. London: Macmillan.

Andrusz, G. D. et al. (eds) (1996) *Cities after Socialism: Urban and Regional Change and Conflict in Post-socialist Societies*. Cambridge, MA: Blackwell.

Bater, J. H. (1980) *The Soviet City: Ideal and Reality*. London: Arnold.

French, R. A. (1995) *Plans, Pragmatism and People: the Legacy of Soviet Planning for Today's Cities*. London: UCL Press.

Pallot, J. and Shaw, D. J. B. (1981) *Planning in the Soviet Union*. London: Croom Helm.

National identity, nationalism and related issues

Bremner, I. and Taras, R. (eds) (1993) *Nations and Politics in the Soviet Successor States*. Cambridge: Cambridge University Press.

Denber, R. (ed.) (1992) *The Soviet Nationality Reader*. Boulder, CO: Westview.

Kaiser, R. (1994) *The Geography of Nationalism in Russia and the USSR*. Princeton, NJ: Princeton University Press.

Lapidus, G. W. (1995) *The New Russia: a Troubled Transformation.* Boulder, CO: Westview.
Saikal, A. and Maley, W. (eds) (1995) *Russia in Search of its Future.* Cambridge: Cambridge University Press.
Sakwa, R. (1996) *Russian Politics and Society,* 2nd edn. London: Routledge.
Simon, G. (1991) *Nationalism and Policy towards the Nationalities in the Soviet Union.* Boulder, CO: Westview.
Smith, G. E. (ed.) (1996) *The Nationalities Question in the Post-Soviet States,* 2nd edn. London: Longman.

Social and demographic geography

DaVanzo, J. (ed.) (1996) *Russia's Demographic 'Crisis'.* Los Angeles: The Rand Corporation.
Desfosses, H. (ed.) (1981) *Soviet Population Policy: Conflicts and Constraints.* New York: Pergamon.
McAuley, A. (1987) *Economic Welfare in the Soviet Union: Poverty, Living Standards and Inequality.* Madison: University of Wisconsin Press.

Russia: regional geography

Bradshaw, M. J. (1996) *Russia's Regions: a Business Analysis.* London: Economist Intelligence Unit.
Hanson, P. (1994) *Regions, Local Power and Economic Change in Russia.* London: RIIA.

Russia, the CIS and the wider world

Aves, J. (1993) *Post-Soviet Transcaucasia.* London: RIIA.
Baranovsky, V. (1997) *Russia and Europe: the Emerging Security Agenda.* Oxford: Oxford University Press.
Central Asia and the Caucasus after the Soviet Union (1994) Gainesville: University of Florida.
Kaiser, R. and Chinn, J. (1996) *Russians as the New Minority: Ethnicity and Nationalism in the Soviet Successor States.* Boulder, CO: Westview Press.
Lewis, R. A. (ed.) (1992) *Geographic Perspectives on Soviet Central Asia.* London: Routledge.
Marples, D. R. (1991) *Ukraine under Perestroika: Ecology, Economics and the Workers' Revolt.* London: Macmillan.
Marples, D. R. (1996) *Belarus: from Soviet Rule to Nuclear Catastrophe.* Macmillan: London.
Neumann, L. B. (1996) *Russia and the Idea of Europe.* London: Routledge.

Odom, W. E. (1995) *Commonwealth or Empire? Russia, Central Asia and the Transcaucasus.* Indianapolis: Hudson Institute.
Smith, G. E. (ed.) (1994) *The Baltic States.* Cambridge: Cambridge University Press.
Ukraine: Human Development Report (1995) Kiev: UNDP.
Webber, M. (1997) *CIS Integration Trends: Russia and the Former Soviet South.* London: RIIA.

Atlases

Brawer, M. (1994) *Atlas of Russia and the Independent Republics.* New York: Simon and Schuster.
Brunet, R., Eckert, D. and Kolossov, V. (eds) (1995) *Atlas de la Russie et des Pays Proches.* Montpellier: Reclus.

Bibliography

Adamesku, A. and Kistanov, V. (1990) Razmeshcheniye proizvoditel'nykh sil i razvitiya narodnogo khozyaystva. *Planovoye khozyaystvo*, 6, 109–14.

Amin, A. (1994) *Post-Fordism: a Reader*. Oxford: Blackwell.

Anderson, M. S. (1978) *Peter the Great*. London: Thames and Hudson.

Anderson, M. S. (1987) *Europe in the Eighteenth Century*, 3rd edn. London: Longman.

Arendt, H. (1958) *The Origins of Totalitarianism*. Cleveland: World Publishing.

Aslund, A. (1996) Reform versus 'rent-seeking' in Russia's economic transformation. *Transition*, 2(2), 12–16.

Bahry, D. (1987) *Outside Moscow: Power, Politics and Budgetary Policy in the Soviet Republics*. New York: Columbia University Press.

Barr, B. M. and Braden, K. E. (1988) *The Disappearing Russian Forest: a Dilemma in Soviet Resource Management*. Totowa, NJ: Rowman and Littlefield.

Bartlett, R. (ed.) (1990) *Land Commune and Peasant Community in Russia: Communal Forms in Imperial and Early Soviet Society*. London: Macmillan.

Bass, I. and Dienes, L. (1993) Defence industry legacies and conversion in the post-Soviet realm. *Post-Soviet Geography*, 34(5), 302–17.

Bassin, M. (1983) The Russian Geographical Society, the Amur epoch and the Great Siberian Expedition, 1855–63. *Annals of the Association of American Geographers*, 73, 240–56.

Bassin, M. (1988) Expansion and colonisation on the eastern frontier: views of Siberia and the Far East in the pre-Petrine period. *Journal of Historical Geography*, 14(1), 3–21.

Bassin, M. (1993) Turner, Solov'ev and the frontier hypothesis: the nationalist signification of open spaces. *Journal of Modern History*, 65(3), 473–511.

Bater, J. H. (1980) *The Soviet City: Ideal and Reality*. London: Edward Arnold.

Bater, J. H. (1986) Some recent perspectives on the Soviet city. *Urban Geography*, 7(1), 93–102.

Baykov, A. (1954) The economic development of Russia. *Economic History Review*, Second Series, 7(2), 137–49.

Bennett, R. J. and Estall, R. C. (eds) (1991) *Global Change and Challenge: Geography for the 1990s*. London: Routledge.

Bond, A. R. and Piepenburg, K. (1990) Land reclamation after surface mining in the USSR: economic, political and legal issues. *Soviet Geography*, 31(5), 332–65.

Bond, A. R. and Sagers, M. J. (1991) Checheno-Ingushetia: background to current unrest. *Soviet Geography*, 32(10), 701–6.

Bond, A. R. and Sagers, M. J. (1992) Some observations on the Russian Federation environmental protection law. *Post-Soviet Geography*, 33(7), 463–74.

Bradshaw, M. J. (1995) Foreign trade and inter-republican relations. In D. J. B. Shaw (ed.), *The Post-Soviet Republics: a Systematic Geography*. Harlow: Longman, pp. 132–50.

Bradshaw, M. J. (1996) *Russia's Regions: a Business Analysis*. London: EIU.

Bradshaw, M. J. and Lynn, N. J. (1996) After the Soviet Union: the post-Soviet states in the world system. *Professional Geographer*, 46(4), 439–49.

Bradshaw, M. J. and Palacin, J. A. (1996) An Atlas of the economic performance of Russia's Regions. Russian Regional Research Group, University of Birmingham, Working Paper No. 2.

Bradshaw, M. J. and Shaw, D. J. B. (eds) (1996) Regional problems during economic transition in Russia: case studies. Russian Regional Research Group, University of Birmingham, Working Paper No. 1.

Burke, A. E. (1956) Influence of man upon nature – the Russian view: a case study. In W. L. Thomas Jr (ed.), *Man's Role in Changing the Face of the Earth*. Chicago: University of Chicago Press, pp. 1036–51.

Bylov, G. V. (1995) *Regional Development in Russia in the 1980s–90s: an Overview of Long-term and Recent Trends*. Moscow: Analytical Department of the President of the Russian Federation, Centre for Regional Studies.

Carrere d'Encausse, H. (1992) *The Great Challenge: Nationalities and the Bolshevik State, 1917–30*. New York: Holmes and Meier.

Castells, M. (1996) The informational economy and the new international division of labour. In S. Daniels and R. Lee (eds), *Exploring Human Geography*. London: Arnold, pp. 98–120.

Clem, R. S. and Craumer, P. R. (1995) The geography of the Russian 1995 parliamentary election: continuity, change and correlates. *Post-Soviet Geography*, 36(10), 587–616.

Clem, R. S. and Craumer, P. R. (1996) Roadmap to victory: Boris Yeltsin and the Russian presidential elections of 1996. *Post-Soviet Geography and Economics*, 37(6), 335–54.

Cliff, T. (1964) *Russia: a Marxist Analysis*. London: Socialist Review Publications.

Cliff, T. (1974) *State Capitalism in Russia*. London: Pluto Press.

Cole, J. P. and Filatochev, I. P. (1992) Some observations on migration within and from the former USSR in the 1990s. *Post-Soviet Geography*, 33(7), 432–53.

Cook, E. C. (1992) Agriculture's role in the Soviet economic crisis. In M. Ellman and V. Kontorovich (eds), *The Disintegration of the Soviet Economic System*. London: Routledge, pp. 193–216.

Cooper, J. (1991) *The Soviet Defence Industry: Conversion and Reform*. London: RIIA/Pinter.

Daniels P. W. and Lever, W. F. (eds) (1996) *The Global Economy in Transition*. Harlow: Longman.

DaVanzo, J. (ed.) (1996) *Russia's Demographic 'Crisis'*. Los Angeles: The Rand Corporation.

Davies, R. W. et al. (ed.) (1994) *The Economic Transformation of the Soviet Union, 1913–45*. Cambridge: Cambridge University Press.

Davis, H. and Scase, R. (1985) *Western Capitalism and State Socialism*. Oxford: Basil Blackwell.

Dienes, L. (1987) *Soviet Asia: Economic Development and National Policy Choices*. Boulder, CO: Westview Press.

Diljas, M. (1966) *The New Class: an Analysis of the Communist System*. London: Allen and Unwin.

Diment, G. and Slezkine, Yu. (1993) *Between Heaven and Hell: the Myth of Siberia in Russian Culture*. New York: St Martin's Press.

Dmitriev, C. (1996) New migration tests Russian immigration policy. *Transition*, 2(13), 56–64.

Dmitrieva, O. (1996) *Regional Development: the USSR and After*. London: UCL Press.

Dukes, P. (1982) *The Making of Russian Absolutism*. Harlow: Longman.

Dunlop, J. (1993) Russia: confronting a loss of empire. In I. Bremner and R. Taras (eds), *Nations and Politics in the Soviet Successor States*. Cambridge: Cambridge University Press, pp. 43–72.

Dyker, D. A. (1992) *Restructuring the Soviet Economy*. London: Routledge.

Ellman, M. (1994) The increase in death and disease under 'katastroika'. *Cambridge Journal of Economics*, 18, 329–55.

Ellman, M. and Kontorovich, V. (1992) Overview. In M. Ellman and V. Kontorovich (eds), *The Disintegration of the Soviet Economic System*. London: Routledge, pp. 1–39.

Feshbach, M. and Friendly, A. (1992) *Ecocide in the USSR: Health and Nature under Siege*. London: Aurum Press.

Field, N. C. (1968) Environmental quality and land productivity: a comparison of the agricultural land base of the USSR and North America. *Canadian Geographer*, 12(1), 1–14.

Filatochev, I. V. and Bradshaw, R. P. (1995) The geographical impact of the Russian privatization program. *Post-Soviet Geography*, 36(6), 371–84.

Fitzpatrick, S. (1982) *The Russian Revolution, 1917–32*. Oxford: Oxford University Press.

Fondahl, G. A. (1995) The status of indigenous peoples in the Russian north. *Post-Soviet Geography*, 36(4), 215–24.

Fortescue, S. (1995) Privatization of large-scale Russian industry. In A. Saikal and W. Maley (eds), *Russia in Search of Its Future*. Cambridge: Cambridge University Press, pp. 85–101.

Frankland, M. (1987) *The Sixth Continent: Russia and Mikhail Gorbachev*. London: Hamish Hamilton.

French, R. A. (1987) Changing spatial patterns in Soviet cities: planning or pragmatism? *Urban Geography*, 8(4), 309–20.

French, R. A. (1995a) Demographic and social problems. In D. J. B. Shaw (ed.), *The Post-Soviet Republics: a Systematic Geography*. Harlow: Longman, pp. 89–100.

French, R. A. (1995b) *Plans, Pragmatism and People: the Legacy of Soviet Planning for Today's Cities*. London: UCL Press.

Galbraith, J. K. (1969) *The New Industrial State*. Harmondsworth: Penguin.

Gattrell, P. (1986) *The Tsarist Economy, 1850–1917*. London: Batsford.

Gdaniec, C. (1997) Reconstruction in Moscow's historic centre: conservation, planning and finance strategies. *Geojournal*, 42(4), 377–84.

Gerschenkron, A. (1962) *Economic Backwardness in Historical Perspective*. Cambridge, MA: Harvard University Press.

Geyer, D. (1987) *Der russische Imperialismus*. Leamington Spa: Berg.

Gibson, J. R. (1976) *Imperial Russia in Frontier America: the Changing Geography of Supply of Russian America, 1784–1867*. Oxford: Oxford University Press.

Gibson, J. R. (1994) Interregional migration in the USSR, 1986–90: a final update. *Canadian Geographer*, 38(1), 54–75.

Godlewska, A. and Smith, N. (eds) (1994) *Geography and Empire*. Oxford: Blackwell.

Gorbachev, M. (1987) *Perestroika*. London: Collins.

Gregory, P. R. and Stuart, R.C. (1994) *Soviet and Post-Soviet Economic Structure and Performance*, 5th edn. New York: HarperCollins.

Gritsai, O. (1997) Economic transformation and local urban restructuring in Moscow. Russian Regional Research Group, University of Birmingham, Working Paper No. 11.

Gur'ianova, M. R. (1996) Society and the rural family. *Russian Social Science Review*, 37(3), 66–76.

Gustafson, T. (1981) *Reform in Soviet Politics: Lessons of Recent Policies on Land and Water*. Cambridge: Cambridge University Press.

Gustafson, T. (1989) *Crisis amid Plenty: the Politics of Soviet Energy under Brezhnev and Gorbachev*. Princeton, NJ: Princeton University Press.

Hanson, P. (1995) Evaluating Russia's regions. Unpublished manuscript, University of Birmingham.

Hanson, P. (1996a) Structural change in the Russian economy. *Transition*, 2, 18–21.

Hanson, P. (1996b) Samara. Unpublished paper, University of Birmingham.

Harris, C. D. (1988) Nikolay Ivanovich Vavilov, 1887–1943. *Soviet Geography*, 29(7), 653–7.

Harris, C. D. (1993a) The new Russian minorities: a statistical overview. *Post-Soviet Geography*, 34(1), 1–27.

Harris, C. D. (1993b) Geographic analysis of non-Russian minorities in Russia and its ethnic homelands. *Post-Soviet Geography*, 34(9), 543–97.

Haub, C. (1994) Population change in the former Soviet Republics. *Population Bulletin*, 49(4), 1–52.

Heleniak, T. (1995) Economic transition and demographic change in Russia, 1989–95. *Post-Soviet Geography*, 36(7), 446–58.

Heleniak, T. (1997) Internal migration in Russia during the economic transition. *Post-Soviet Geography and Economics*, 38(2), 81–104.

Hellie, R. (1971) *Enserfement and Military Change in Muscovy*. Chicago: Chicago University Press.

Hettne, B. (1990) *Development Theory and the Three Worlds*. Harlow: Longman.

Hewett, E. A. (1988) *Reforming the Soviet Economy: Equality versus Efficiency*. Washington, DC: Brookings Institution.

Hilton, A. C. (1996) Lost opportunities at the CIS migration conference. *Transition*, 2(13), 52–4.

Holmes, L. (1995) Russia's relations with the former external empire. In A. Saikal and W. Maley (eds), *Russia in Search of Its Future*. Cambridge: Cambridge University Press, pp. 123–41.

Hooson, D. J. M. (1959) Some recent developments in the content and theory of Soviet geography. *Annals of the Association of American Geographers*, 49, 73–82.

Hooson, D. J. M. (1964) *A New Soviet Heartland*. Princeton, NJ: Van Nostrand.

Hosking, G. (1992) *A History of the Soviet Union, 1917–1991*, final edition. London: Fontana.

Hughes, J. (1996) Moscow's bilateral treaties add to confusion. *Transition*, 20 September, 39–43.

Hunczak, T. (ed.) (1974) *Russian Imperialism from Ivan the Great to the Revolution*. New Brunswick, NJ: Rutgers University Press.

Ioffe, G. V., Nefedova, T. G. and Runova, T. G. (1989) Intensification of agriculture in the European USSR: regional aspects. *Soviet Geography*, 30(1), 49–64.

Kaiser, R. (1991) Nationalism: the challenge to Soviet federalism. In M. J. Bradshaw (ed.), *The Soviet Union: a New Regional Geography*. London: Belhaven Press, pp. 39–65.

Kaiser, R. (1994) *The Geography of Nationalism in Russia and the USSR*. Princeton, NJ: Princeton University Press.

Kaiser, R. and Chinn, J. (1995) Russian–Kazakh relations in Kazakhstan. *Post-Soviet Geography*, 36(5), 257–73.

Kaiser, R. and Chinn, J. (1996) *Russians as the New Minority: Ethnicity and Nationalism in the Soviet Successor States*. Boulder, CO: Westview Press.

Keep, J. (1972) Shade and light in the history of Russian administration. *Canadian Slavic Studies*, 6, 1–9.

Keep, J. (1995) *Last of the Empires: a History of the Soviet Union, 1945–1991*. Oxford: Oxford University Press.

Kerner, R. J. (1946) *The Urge to the Sea: the Course of Russian History*. Berkeley and Los Angeles: University of California Press.

Kirkow, P. (1997) Transition in Russia's principal coastal gateways. *Post-Soviet Geography and Economics*, 38(5), 296–314.

Kirkow, P. and Hanson, P. (1994) The potential for autonomous regional development in Russia: the case of Primorskiy Kray. *Post-Soviet Geography*, 35(2), 63–88.

Klyuchevsky, V. O. (1937) *Kurs russkoy istorii, volume 1*. Moscow.

Knox, P. and Agnew, J. (1989) *The Geography of the World Economy*. London: Edward Arnold.

Kochan, L. and Abraham, R. (1983) *The Making of Modern Russia*. Harmondsworth: Penguin.

Kochurov, B. (1995) European Russia. In P. R. Pryde (ed.), *Environmental Resources and Constraints in the Former Soviet Republics*. Boulder, CO: Westview Press, pp. 41–59.

Kohn, H. (ed.) (1962) *The Mind of Modern Russia*. New York: Harper and Row.

Kolstoe, P. (1995) *Russians in the Former Soviet Republics*. London: Hurst.

Konstitutsiya Rossiyskoy Federatsii (1997) Moscow: Izdatel'stvo 'Spark'.

Kontorovich, V. (1992) The railroads. In M. Ellman and V. Kontorovich (eds), *The Disintegration of the Soviet Economic System*. London: Routledge, pp. 174–92.

Kontorovich, V. (1996) Imperial legacy and the transformation of the Russian economy. *Transition*, 17, 23 August, 22–25, 64.

Kuznetsov, N. T. and L'vovich, M. I. (1971) Water. In I. P. Gerasimov, D. M. Armand and K. M. Yefron (eds), *Natural Resources of the Soviet Union: Their Use and Renewal*. San Francisco: W. H. Freeman, pp. 11–39.

Lane, D. (1982) *The End of Social Inequality*. London: Allen and Unwin.

Lash, S. and Urry, J. (1987) *The End of Organized Capitalism*. Cambridge: Polity.

Lerman Z., Tankhilevich, Ye., Mozhin, K. and Sapova, N. (1994) Self-sustainability of subsidiary household plots: lessons for privatization of agriculture in former socialist countries. *Post-Soviet Geography*, 35(9), 526–42.

Lewin, M. (1988) *The Gorbachev Phenomenon: an Historical Interpretation*. London: Radius.

Liebowitz, R. (1987) Soviet investment strategy: a further test of the 'equalization hypothesis'. *Annals of the Association of American Geographers*, 77(3), 396–407.

Liebowitz, R. (1991) Spatial inequality under Gorbachev. In M. J. Bradshaw (ed.), *The Soviet Union: a New Regional Geography*, London: Belhaven Press, pp. 15–37.

Linge, G. J. R., Karaska, G. J. and Hamilton, F. E. I. (1978) Appraisal of the Soviet TPC concept. *Soviet Geography*, 19(10), 681–97.

Lloyd, P. E. and Dicken, P. (1977) *Location in Space: a Theoretical Approach to Economic Geography*, 2nd edn. London: Harper and Row.

Lowenthal, D. and Prince, H. C. (1965) English landscape tastes. *Geographical Review*, 55, 186–222.

Lydolph, P. E. (1977) *World Survey of Climatology, volume 7. Climates of the Soviet Union*. Amsterdam: Elsevier.

Lynn, N. J. (1996) A political geography of the republics of the Russian Federation. PhD thesis, University of Birmingham.

Mackinder, H. J. (1904) The geographical pivot of history. *Geographical Journal*, 23, 421–37.

Mackinder, H. J. (1919) *Democratic Ideals and Reality: a Study in the Politics of Reconstruction*. London: Constable.

Magnusson, M.-L. (1996) Relations between the Federal centre and the subjects of the Russian Federation. In L. B. Wallin (ed.), *Lectures and Contributions to East European Studies at FOA*. Stockholm: FOA, pp. 33–42.

Maley, W. (1995) The shape pf the Russian macroeconomy. In A. Saikal and W. Maley (eds), *Russia in Search of its Future*. Cambridge: Cambridge University Press, pp. 48–65.

Marples, D. R. (1993) The post-Soviet nuclear power program. *Post-Soviet Geography*, 34(3), 172–84.

Massey, D. (1984) *Spatial Divisions of Labour: Social Structures and the Geography of Production*. Basingstoke: Macmillan.

Mazurkiewicz, L. (1992) *Human Geography in Eastern Europe and the former Soviet Union*. London: Belhaven Press.

Medish, V. (1980) Special status of the RSFSR. In E. Allworth (ed.), *Ethnic Russia in the USSR: The Dilemma of Dominance*. New York: Pergamon, pp. 188–96.

Meinig, D. W. (1968) The macrogeography of Western imperialism. In F. H. Gale and G. H. Lawton (eds), *Settlement and Encounter*. Oxford: Oxford University Press, pp. 213–40.

Meinig, D. W. (1979) Symbolic landscapes. In J. B. Jackson et al. (eds), *The Interpretation of Ordinary Landscapes*, Oxford: Oxford University Press, pp. 164–92.

Mendel, A. P. (1961) *Dilemmas of Progress in Tsarist Russia: Legal Marxism and Legal Populism*. Cambridge, MA: Harvard University Press.

Micklin, P. (1986) The status of the Soviet Union's north–south water transfer projects before their abandonment in 1985–6. *Soviet Geography*, 27(5), 287–329.

Micklin, P. (ed.) (1992) Special issue on the Aral Sea crisis. *Post-Soviet Geography*, 33(5), 269–331.

Miller, J. (1993) *Mikhail Gorbachev and the End of Soviet Power*. London: Macmillan.

Monroe, S. D. (1992) Chelyabinsk: the evolution of disaster. *Post-Soviet Geography*, 33(8), 533–45.

Morvant, P. (1995) Alarm over falling life expectancy. *Transition*, 1(19), 40–5.

Morvant, P. and Rutland, P. (1996) Russian workers face the market. *Transition*, 3(13), 6–11.

Moskovskiy stolichnyy region (1988): territorial'naya struktura i prirodnaya sreda. Moscow: IGAN SSSR.

Munting, R. (1982) *The Economic Development of the USSR*. London: Croom Helm.

Murray, R. (1988) Life after Henry (Ford). *Marxism Today*, October, 8–13.

Natsional'nyy sostav naseleniya SSSR (1991) Moscow: Finansy i tatistika.

Nefedova, T. and Treyvish, A. (1994) *Rayony Rossii i drugikh yevropeyskikh stran s perekhodnoy ekonomikoy v nachele 90-kh*. Moscow: IGAN RAN.

Noren, J. H. (1994) The Russian military industrial sector and conversion. *Post-Soviet Geography*, 35(9), 495–521.

North, R. N. (1995) Transport. In D. J. B. Shaw (ed.), *The Post-Soviet Republics: a Systematic Geography*. Harlow: Longman, pp. 66–88.

North, R. N. (1997) Transport in a new reality. In M. J. Bradshaw (ed.), *Geography and Transition in the Post-Soviet Republics*. London: Wiley, pp. 209–27.

Nove, A. (1972) *An Economic History of the USSR*. Harmondsworth: Penguin.

Nove, A. (1987) *The Soviet Economic System*, 3rd edn. London: Allen and Unwin.

Novikov, A. (1997) Between space and race: rediscovering Russian cultural geography. In M. J. Bradshaw (ed.), *Geography and Transition in the Post-Soviet Republics*. London: Wiley, pp. 43–57.

OECD (1995) *OECD Economic Surveys: the Russian Federation*. Paris: OECD.

OECD (1996) *Environmental Information Systems in the Russian Federation: an OECD Assessment*. Paris: OECD.

Ofer, G. (1976) Industrial structure, urbanization and the growth strategy of socialist countries. *Quarterly Journal of Economics*, 90, May, 219–44.

O'Loughlin, J., Shin, M. and Talbot, P. (1996) Political geographies and cleavages in the Russian parliamentary elections. *Post-Soviet Geography and Economics*, 37(6), 355–85.

Osherenko, G. (1995) Indigenous political and property rights and economic/ environmental reform in north-west Siberia. *Post-Soviet Geography*, 36(4), 225–37.

Paddison, R. (1983) *The Fragmented State: the Political Geography of Power*. Oxford: Blackwell.

Pallot, J. (1979) Rural settlement planning in the USSR. *Soviet Studies*, 31(2), 214–30.

Pallot, J. (1990) Rural depopulation and the restoration of the Russian village under Gorbachev. *Soviet Studies*, 42(4), 655–74.

Pallot, J. and Shaw, D. J. B. (1981) *Planning in the Soviet Union*. London: Croom Helm.

Pallot, J. and Shaw, D. J. B. (1990) *Landscape and Settlement in Romanov Russia, 1613–1917*. Oxford: Clarendon Press.

Panel on the State of the Soviet Environment at the Start of the Nineties (1990) *Post-Soviet Geography*, 31(6), 401–68.

Parkins, M. F. (1953) *City Planning in Soviet Russia*. Chicago: University of Chicago Press.

Peet, R. (1989) Introduction to new models of uneven development and regional change. In R. Peet and N. Thrift (eds), *New Models in Geography: the Political Economy Perspective, volume 1*. London: Unwin Hyman, pp. 105–13.

Peterson, D. P. (1993) *Troubled Lands: the Legacy of Soviet Environmental Destruction*. Boulder, CO: Westview.

Peterson, D. P. (1995a) Building bureaucratic capacity in Russia: federal and regional responses to the post-Soviet environmental challenge. In J. DeBardeleben and J. Hannigan (eds), *Environmental Security and Quality after Communism*. Boulder, CO: Westview, pp. 107–26.

Peterson, D. P. (1995b) Russia's environment and natural resources in light of economic regionalization. *Post-Soviet Geography*, 36(5), 291–309.

Pivovarov, Yu. V. (1992) Urbanizatsiya v SSSR: makroregional'nyye razlichiya, stadial'no-regional'nyy podkhod, kontrastnost' rasseleniya. *Izvestiya Akademii Nauk: seriya geograficheskaya*, 1, 52–63.

Pratt, M. L. (1992) *Imperial Eyes: Travel Writing and Transculturation.* London: Routledge.

Pryde, P. R. (1972) *Conservation in the Soviet Union.* Cambridge: Cambridge University Press.

Pryde, P. R. (1991) *Environmental Management in the Soviet Union.* Cambridge: Cambridge University Press.

Pryde, P. R. (1994) Observations on the mapping of critical environmental zones in the former Soviet Union. *Post-Soviet Geography*, 35(1), 38–49.

Pryde, P. R. (1995) Russia: an overview of the Federation. In P. R. Pryde (ed.), *Environmental Resources and Constraints in the Former Soviet Republics.* Boulder, CO: Westview Press, pp. 25–39.

Pryde, P. R. and Bradley, D. J. (1994) The geography of radioactive contamination in the former USSR. *Post-Soviet Geography*, 35(10), 557–93.

Radvanyi, J. (1992) And what if Russia breaks up? Towards new regional subdivisions. *Post-Soviet Geography*, 33(2), 69–77.

Raeff, M. (1971) Patterns of Russian imperial policy toward the nationalities. In E. Allworth (ed.), *Soviet Nationality Problems.* New York: Columbia University Press, pp. 22–42.

Rahr, A. (1992) 'Atlanticists' versus 'Eurasians' in Russian foreign policy. *RFE/RL Research Report*, 1(22), 17–22.

Rossiya v tsifrakh. Moscow: Finansy i statistika, various years.

Rossiyskiy statisticheskiy yezhegodnik. Moscow: Goskomstat, various years.

Rowland, R. H. (1994) Declining towns in the former USSR. *Post-Soviet Geography*, 35(6), 352–65.

Rowland, R. H. (1995) Declining towns in Russia, 1989–93. *Post-Soviet Geography*, 36(7), 436–45.

Rowland, R. H. (1996a) Russia's disappearing towns: new evidence of urban decline, 1979–94. *Post-Soviet Geography*, 37(2), 63–87.

Rowland, R. H. (1996b): Russia's secret cities. *Post-Soviet Geography*, 37(7), 426–62.

Rutland, P. (1996a) Firms trapped between the past and the future. *Transition*, 2(6), 26–32.

Rutland, P. (1996b) Russia's energy empire under strain. *Transition*, 2(9), 6–11.

Rywkin, M. (ed.) (1988) *Russian Colonial Expansion to 1917.* London: Mansell.

Sagers, M. J. (1990) Review of energy industries in 1989. *Soviet Geography*, 31(4), 278–314.

Sagers, M. J. (1995a) The Russian natural gas industry in the mid-1990s. *Post-Soviet Geography*, 36(9), 521–64.

Sagers, M. J. (1995b) Prospects for oil and gas development in Russia's Sakhalin Oblast. *Post-Soviet Geography*, 36(5), 274–90.

Sagers, M. J. (1996) Russian crude oil production in 1996: conditions and prospects. *Post-Soviet Geography and Economics*, 37(9), 523–87.

Sakwa, R. (1989) *Soviet Politics: an Introduction.* London: Routledge.

Sakwa, R. (1990) *Gorbachev and His Reforms, 1985–90.* New York: Philip Allan.

Sanches-Andres, A. (1995) The transformation of the Russian defence industry. *Europe–Asia Studies,* 47(8), 1269–92.

Saushkin, Yu. G. (1973) *Ekonomicheskaya geografiya: istoriya, teoriya, metody, praktika.* Moscow: Mysl'.

Sayer, A. (1989) Post-Fordism in question. *International Journal of Urban and Regional Research,* 13(4), 666–95.

Scherbakova, A. and Monroe, S. (1995) The Urals and Siberia. In P. R. Pryde (ed.), *Environmental Resources and Constraints in the Former Soviet Republics.* Boulder, CO: Westview Press, pp. 61–77.

Schiffer, J. (1989) *Soviet Regional Economic Policy: the East–West Debate over Pacific Siberian Development.* London: Macmillan.

Schmidt, A. (1971) *The Concept of Nature in Marx.* London: NLB.

Semenov-Tyan-Shanskiy, V. P. (ed.) (1899–1914) *Rossiya: polnoye geograficheskoye opisaniye nashego otechestva,* 11 volumes. St Petersburg: A. F. Devrien.

Senghaas, D. (1985) *The European Experience.* Leamington Spa: Berg Publishers.

Seton-Watson, H. (1967) *The Russian Empire, 1801–1917.* Oxford: Oxford University Press.

Shanin, T. (1985) *Russia as a Developing Society.* London: Macmillan.

Shaw, D. J. B. (1983) Southern frontiers of Muscovy, 1550–1700. In J. H. Bater and R. A. French (eds), *Studies in Russian Historical Geography, volume 1.* London: Academic Press, pp. 117–42.

Shaw, D. J. B. (1985) Spatial dimensions in Soviet central planning. *Transactions of the Institute of British Geographers,* n.s., 10, 401–12.

Shaw, D. J. B. (1986) Regional Planning in the USSR. *Soviet Geography,* 27(7), 469–84.

Shaw, D. J. B. (1989) The settlement of European Russia during the Romanov period. *Soviet Geography,* 30(3), 207–28.

Shaw, D. J. B. (1992) Russian Federation treaty signed. *Post-Soviet Geography,* 33(6), 414–17.

Shaw, D. J. B. (1993) Geographic and historical observations on the future of a federal Russia. *Post-Soviet Geography,* 34(8), 530–40.

Shaw, D. J. B. (1995) Fifteen successor states: fifteen and more futures? In D. J. B. Shaw (ed.), *The Post-Soviet Republics: a Systematic Geography.* London: Longman, pp. 151–65.

Simon, G. (1991) *Nationalism and Policy towards the Nationalities in the Soviet Union.* Boulder, CO: Westview.

Skilling, H. G. and Griffiths, F. (eds) (1971) *Interest Groups in Soviet Politics.* Princeton, NJ: Princeton University Press.

Smith, A. D. (1991) *National Identity.* Harmondsworth: Penguin.

Smith, G. E. (1985) Ethnic nationalism in the Soviet Union: territory, cleavage and control. *Environment and Planning C. Government and Policy,* 3, 49–73.

Smith, G. E. (1989) Gorbachev's greatest challenge: perestroyka and the national question. *Political Geography Quarterly,* 8(1), 7–20.

Smith, G. E. (ed.) (1995) *Federalism.* London: Longman.

Solzhenitsyn, A. (1991) *Rebuilding Russia*. London: Harvill.

Spechler, M. C. (1980) The regional concentration of industry in Imperial Russia. *Journal of European Economic History*, 9(2), 401–29.

Sumner, B. H. (1944) *Survey of Russian History*. London: Methuen.

Sutherland D. and Hanson, P. (1996) Structural change in the economies of Russia's regions. *Europe–Asia Studies*, 48(3), 367–92.

Talbott, S. (1974) *Khrushchev Remembers*. London: Andre Deutsch.

Taylor, P. (1993) *Political Geography: World Economy, Nation State and Locality*, 3rd edn. Harlow: Longman.

Thrift, N. and Forbes, D. (1986) *The Price of War: Urbanization in Vietnam, 1954–85*. London: Allen and Unwin.

Todres, V. (1995) Bashkortostan seeks sovereignty – step by step. *Transition*, 7, 56–9.

Treyvish, A. I., Pandit, K. K. and Bond, A. R. (1993) Macrostructural employment shifts and urbanization in the former USSR: an international perspective. *Post-Soviet Geography*, 34(3), 157–71.

Tuminez, A. S. (1996) Nationalism and the interest in Russian foreign policy. In C. A. Wallander (ed.), *The Sources of Russian Foreign Policy after the Cold War*. Boulder, CO: Westview Press, pp. 41–68.

Van Atta, D. (1994) Agrarian reform in post-Soviet Russia. *Post-Soviet Affairs*, 10(2), 159–90.

Vanderheide, D. (1980) Ethnic significance of the non-Black Earth renovation project. In E. Allworth (ed.), *Ethnic Russia in the USSR: the Dilemma of Dominance*. New York: Pergamon, pp. 218–28.

Vitebsky, P. (1996) The northern minorities. In G. Smith (ed.), *The Nationalities Question in the Post-Soviet States*. Harlow: Longman, pp. 94–112.

von Herberstein, S. (1966) *Description of Moscow and Muscovy, 1557*. London: J. M. Dent.

Walicki, A. (1969) *A History of Russian Thought from the Enlightenment to Marxism*. Oxford: Clarendon Press.

Walker, E. W. (1992) The new Russian constitution and the future of the Russian Federation. *The Harriman Institute Forum*, 5(10).

Wallerstein, I. (1974) *The Modern World System: Capitalist Agriculture and the Origins of the World Economy in the Sixteenth Century*. New York: Academic Press.

Wallerstein, I. (1979) *The Capitalist World Economy*. Cambridge: Cambridge University Press.

Wallerstein, I. (1980) *The Modern World System: Mercantilism and the Consolidation of the European World Economy, 1600–1750*. New York: Academic Press.

Weiner, D. (1988) *Models of Nature: Ecology, Conservation and Cultural Revolution in Soviet Russia*. Bloomington: Indiana University Press.

Wheare, K. C. (1963) *Federal Government*, 4th edn. Oxford: Oxford University Press.

White, C. (1987) *Russia and America: the Roots of Economic Divergence*. London: Croom Helm.

White, S. (1993) *After Gorbachev*. Cambridge: Cambridge University Press.

Zaslavsky V. (1982) *The Neo-Stalinist State: Class, Ethnicity and Consensus in Soviet Society*. Brighton: Harvester.

Zaslavsky, V. (1993) Success and collapse: traditional Soviet nationality policy. In I. Bremner and R. Taras (eds), *Nations and Politics in the Soviet Successor States*. Cambridge: Cambridge University Press, pp. 29–42.

Zubov, A. B. (1994) The Soviet Union: from an empire into nothing? *Russian Social Science Review*, 32, 37–67.

312 Index